In Vitro Cultivation of Micro-organisms

BOOKS IN THE BIOTOL SERIES

BIOTECHNOLOGY BY OPEN LEARNING

In Vitro Cultivation of Micro-organisms

PUBLISHED ON BEHALF OF :

Open universiteit   and   **Thames Polytechnic**

Valkenburgerweg 167
6401 DL Heerlen
Nederland

Avery Hill Road
Eltham, London SE9 2HB
United Kingdom

Butterworth-Heinemann Ltd
Linacre House, Jordan Hill, Oxford OX2 8DP

 PART OF REED INTERNATIONAL BOOKS

OXFORD   LONDON   BOSTON
MUNICH   NEW DELHI   SINGAPORE   SYDNEY
TOKYO   TORONTO   WELLINGTON

First published 1992

British Library Cataloguing in Publication Data
A catalogue record for this book is
available from the British Library

Library of Congress Cataloguing in Publication Data
A catalogue record for this book is
available from the Library of Congress

ISBN 0 7506 0507 3

Composition by Thames Polytechnic
Printed and Bound in Great Britain by
Thomson Litho, East Kilbride, Scotland

The Biotol Project

The BIOTOL team

**OPEN UNIVERSITEIT,
THE NETHERLANDS**
Dr M. C. E. van Dam-Mieras
Professor W. H. de Jeu
Professor J. de Vries

**THAMES POLYTECHNIC,
UK**
Professor B. R. Currell
Dr J. W. James
Dr C. K. Leach
Mr R. A. Patmore

This series of books has been developed through a collaboration between the Open universiteit of the Netherlands and Thames Polytechnic to provide a whole library of advanced level flexible learning materials including books, computer and video programmes. The series will be of particular value to those working in the chemical, pharmaceutical, health care, food and drinks, agriculture, and environmental, manufacturing and service industries. These industries will be increasingly faced with training problems as the use of biologically based techniques replaces or enhances chemical ones or indeed allows the development of products previously impossible.

The BIOTOL books may be studied privately, but specifically they provide a cost-effective major resource for in-house company training and are the basis for a wider range of courses (open, distance or traditional) from universities which, with practical and tutorial support, lead to recognised qualifications. There is a developing network of institutions throughout Europe to offer tutorial and practical support and courses based on BIOTOL both for those newly entering the field of biotechnology and for graduates looking for more advanced training. BIOTOL is for any one wishing to know about and use the principles and techniques of modern biotechnology whether they are technicians needing further education, new graduates wishing to extend their knowledge, mature staff faced with changing work or a new career, managers unfamiliar with the new technology or those returning to work after a career break.

Our learning texts, written in an informal and friendly style, embody the best characteristics of both open and distance learning to provide a flexible resource for individuals, training organisations, polytechnics and universities, and professional bodies. The content of each book has been carefully worked out between teachers and industry to lead students through a programme of work so that they may achieve clearly stated learning objectives. There are activities and exercises throughout the books, and self assessment questions that allow students to check their own progress and receive any necessary remedial help.

The books, within the series, are modular allowing students to select their own entry point depending on their knowledge and previous experience. These texts therefore remove the necessity for students to attend institution based lectures at specific times and places, bringing a new freedom to study their chosen subject at the time they need and a pace and place to suit them. This same freedom is highly beneficial to industry since staff can receive training without spending significant periods away from the workplace attending lectures and courses, and without altering work patterns.

Contributors

AUTHORS

Dr T.G. Cartledge, Nottingham Polytechnic, Nottingham, UK

Dr J.S. Drijver-de Haas, Open universiteit, Heerlen, The Netherlands

Dr R.O. Jenkins, Leicester Polytechnic, Leicester, UK

Dr E.J. Middelbeek, Open universiteit, Heerlen, The Netherlands

EDITOR

Dr T.G. Cartledge, Nottingham Polytechnic, Nottingham, UK

SCIENTIFIC AND COURSE ADVISORS

Dr M.C.E. van Dam-Mieras, Open universiteit, Heerlen, The Netherlands

Dr C.K. Leach, Leicester Polytechnic, Leicester, UK

ACKNOWLEDGEMENTS

Grateful thanks are extended, not only to the authors, editors and course advisors, but to all those who have contributed to the development and production of this book. They include Mrs A. Allwright, Miss K. Brown, Mrs N. Cartledge, Miss J. Skelton and Professor R. Spier. The development of this BIOTOL text has been funded by COMETT, The European Community Action programme for Education and Training for Technology, by the Open universiteit of The Netherlands and by Thames Polytechnic.

Project Manager: Dr J.W. James

Contents

How to use an open learning text

An open learning text presents to you a very carefully thought out programme of study to achieve stated learning objectives, just as a lecturer does. Rather than just listening to a lecture once, and trying to make notes at the same time, you can with a BIOTOL text study it at your own pace, go back over bits you are unsure about and study wherever you choose. Of great importance are the self assessment questions (SAQs) which challenge your understanding and progress and the responses which provide some help if you have had difficulty. These SAQs are carefully thought out to check that you are indeed achieving the set objectives and therefore are a very important part of your study. Every so often in the text you will find the symbol Π, our open door to learning, which indicates an activity for you to do. You will probably find that this participation is a great help to learning so it is important not to skip it.

Whilst you can, as a open learner, study where and when you want, do try to find a place where you can work without disturbance. Most students aim to study a certain number of hours each day or each weekend. If you decide to study for several hours at once, take short breaks of five to ten minutes regularly as it helps to maintain a higher level of overall concentration.

Before you begin a detailed reading of the text, familiarise yourself with the general layout of the material. Have a look at the contents of the various chapters and flip through the pages to get a general impression of the way the subject is dealt with. Forget the old taboo of not writing in books. There is room for your comments, notes and answers; use it and make the book your own personal study record for future revision and reference.

At intervals you will find a summary and list of objectives. The summary will emphasise the important points covered by the material that you have read and the objectives will give you a check list of the things you should then be able to achieve. There are notes in the left hand margin, to help orientate you and emphasise new and important messages.

BIOTOL will be used by universities, polytechnics and colleges as well as industrial training organisations and professional bodies. The texts will form a basis for flexible courses of all types leading to certificates, diplomas and degrees often through credit accumulation and transfer arrangements. In future there will be additional resources available including videos and computer based training programmes.

Preface

To many, the term biotechnology conjures up a picture of large volume reactors in which cultures of micro-organisms are used to make products such as wine, beer and antibiotics. Of course, biotechnology is much more diverse than this, but the common misconception reflects the fact that the use of micro-organisms dominates many biotechnological endeavours. There are several reasons for this. The fast growth rates of micro-organisms and the associated high rates of metabolism coupled to their metabolic diversity ensure that micro-organisms are often the agents of choice. These features together with the relative ease by which they may be genetically manipulated and increasingly detailed knowledge of their metabolism provides further impetus for their systematic exploitation to produce an enormous array of metabolites and enzymes. Opportunities to use micro-organisms to deal with domestic and industrial waste and contributors to agriculture are being increasingly grasped.

Within the context of the BIOTOL series, the text dealing with the *in vitro* cultivation of micro-organisms needs no further justification. The ability to cultivate micro-organisms underpins so many biotechnological activities ranging from the small volumes enclosed in the genetic manipulation of plant and animal systems to the enormous capacity of many large volume processes. This text aims to provide the essential knowledge of the core processes involved in the cultivation of micro-organism irrespective of the organism or the scale of the operation. Discussion of the *in vitro* cultivation of cells from higher plants and animals has been specifically excluded. Although cultivation of these cells has much in common with the cultivation of micro-organisms, they do display some important differences and two BIOTOL texts have been produced to cover these important groups ('In vitro Culitvation of Plant Cells' and 'In vitro Cultivation of Animals Cells').

A key feature of any process involving micro-organisms is the need for containment. Some micro-organisms are pathogenic and need to be prevented from either entering an industrial process or for that matter, leaving a process to infect workers or the community at large. Many processes need to be conducted with pure cultures, contamination by undesirable strains leads to poorer, unpredictable productivity. This the need to prevent the unwanted transfer of micro-organisms is essential. Thus aspect of *in vitro* cultivation of micro-organisms is introduced in the first chapter and is a recurrent theme within the text.

The major part of the text deals with the nutrition of micro-organisms with particular emphasis on media design, the evaluation and characterisation of growth and the influence and control of the physical and chemical parameters which influence microbial performance in culture. A chapter is devoted to discussion of the cultivation of viruses, especially bacteriophages. The importance of bacteriophages in contemporary biotechnology lies in their use as genetic vectors and they also offer some potential as antibacterial agents for some bacterial infections. The final chapter deals with the chemical agents that may be used to control microbial growth. Disinfection and disinfection policies are key components of good microbiological practice at both laboratory and manufacturing scales of operation. Good microbiological practice is in its turn, essential to good laboratory and good manufacturing practice.

Although a discussion of growth in batch and continuous culture is included, this discussion focusses predominantly on the biological and scientific issues involved. For those seeking to expand their knowledge into the engineering issues of *in vitro* cultivation of micro-organisms, we recommend the BIOTOL series of technology texts, especially 'Bioprocess Technology: Modelling and Transport Phenomena'; 'Operational Modes of Bioreactors' and 'Bioreactor Design and Product Yield'.

The authors are to be congratulated on their synthesis of a sound and logical development of this topic. The quality and relevance of the technical material they have used are matched by their ability to design interactive activities within the text which aid the reader towards a full understanding of the issues under discussion. We encourage readers to take full advantage of these opportunities.

Scientific and Course Advisors: Dr M.C.E. van Dam-Mieras
Dr C.K. Leach

Introduction to microbial growth and cultivation

Introduction to microbial growth and cultivation

1.1 Introduction

Microbiology is a very specialised area within the science of biology and the strategies and methods required to grow micro-organisms are very different to those required for growth of, for example plants. The chapters of this text are structured and ordered such that the reader can follow the theme of growth of mainly unicellular organisms which divide regularly by asexual means forming two daughter cells from each parent cell. The major exception to this system is the group of viruses and they will be treated separately.

We will not include discussion of animal and plant cells in culture within this test. Although cultivation techniques for these systems have much in common with the cultivation of micro-organisms, they also have many special features. A description of the cultivation of plant and animal cells is given in two other BIOTOL texts: 'In vitro Cultivation of Plant Cells' and 'In vitro Cultivation of Animal Cells'.

It is appropriate at this stage to introduce a few words of caution. Throughout the book we will mention the fact that micro-organisms, can particularly in high numbers, be dangerous and should always be treated with caution and great respect. This text will provide you with insight into the theoretical and practical aspects of microbial cell culture which you could then employ under supervision at your place of work, if suitable, or in an appropriate laboratory under qualified supervision. The text is not intended to be a 'do-it-yourself' manual encouraging you to experiment in, for example your own home. To do so could be both dangerous and quite unwise.

1.1.1 The structure of the text

Chapter 1 introduces the microbial world and gives you an overview of the history of microbiology followed by a discussion of appropriate current safety legislation and recommendations.

Chapter 2 gives an overview of the chemical composition of the cell leading on to the nutritional requirements of different types of micro-organisms and thus the design of laboratory media. The types of media, culturing conditions and sterilisation methods are discussed and you are encouraged to investigate the strategy behind selecting suitable source(s) of micro-organisms and obtaining pure cultures.

Chapter 3 describes the estimation of biomass, by cell numbers, weight or volume. Several methods of measurement are examined and a critical evaluation of the advantages, disadvantages and uses of each method are described.

Chapter 4 examines the growth of micro-organisms in the various forms of batch culture and introduces you into the energetics of growth.

Chapter 5 discusses the environmental factors which influence growth indicating how we predict the effect on growth that changes in environmental parameters would bring about. Finally in this chapter we explain how these factors may be manipulated to our advantage.

Chapter 6 examines growth of micro-organisms in continuous culture. This system is compared to batch cultures and suitable treatment of the topic involves discussion and derivation of the mathematical relationships of various parameters and terms relevant to microbial cell cultivation. Industrial uses and applications of continuous culture are also discussed.

Chapter 7 examines the influence of and the control of selected factors which affect growth in a chemostat, for example pH, temperature and oxygen concentration.

Chapter 8 deals exclusively with viruses beginning with a brief history of the science of virology and followed by an examination of viral structure and classification. The forms of viral replication are discussed with particular emphasis on bacterial viruses (bacteriophages) and the methods available for production of large quantities of viruses in the laboratory. Methods available for estimating viral numbers are also described and evaluated.

In the final chapter, the chemical control of growth is examined. The distinction between antiseptic, disinfectant, metabolic inhibitor and antibiotics is explained and the kinetics of microbial death is described.

1.2 The composition and characteristics of the microbial world

definition of
microbiology

A useful, working definition of microbiology is the study of organisms too small to be seen by the naked eye. The human eye can clearly see objects which are 1 mm or more in size but objects which are of the order of 0.2 mm or less cannot be resolved. Thus we could defines microbiology as the study of organisms (or more commonly - micro-organisms) which are less than 1 mm in size. We will see in later discussion that this is not a strict definition and in practice this definition is not a rigid one.

The millimetre or mm is one thousandth of a meter but within microbiology it is easier to denote length using a millionth (10^6) metre which is more commonly known as a micrometre (μm) or micron (μ). The length of a typical bacterium as we shall see is commonly about 2 μm.

It has been known for a long time that all living organisms are composed of one or more fundamental units called cells. Thus if we were looking at any organism at the individual cell level the study could, in one sense, be termed microbiology. However, we usually reserve the term for living systems in which the whole organism, that is the entity which can and normally exists independently, is too small to be seen by the unaided eye.

∏   Write a list of the groups of living organisms which normally exist independently as unicells?

Some of the major groups which you may have considered include the bacteria, fungi, algae and protozoa. Possible your answer would have included viruses; we will refer to these again shortly.

The terms microbiology and micro-organisms are really very ill defines and have no taxonomic significance. From the organisms listed above, the bacteria, protozoa and viruses can exist independently as single units though the first two sometimes purposely produce aggregates. Fungi and algae, however, vary from small, simple unicellular structures through to multicellular structures, for example mushrooms and seaweed. The study of mushrooms several centimetres high and of seaweed several metres long is nor strictly speaking microbiology, although because of their relationship with unicellular forms they are frequently included in courses on microbiology.

To allow us to investigate the characteristics of micro-organisms it is necessary to return to the concept of the cell. The ability to successfully exist independently - the concept of unicellularity - is an important one not displayed by higher organisms. Unicellular organisms, sometimes called 'unicells', are generally relatively simple cells having a high degree of adaptability. Early theories on classification or the assigning of living systems into groups suggested that all living matter was either 'plant' or 'animal'. As our knowledge of microbiology increased during the last century micro-organisms were themselves assigned to plants or animals largely on the basis of the presence of motility (indicating an animal cell) or photosynthesis (indicating a plant cell). From the beginning of this century, however, it became apparent that there were many micro-organisms which could not fit into one of the two categories above. For example, there are non-motile, non-photosynthetic protozoa and also motile, photosynthetic protozoa.

prokaryotic cells

eukaryotic cells

Around 1950 the development of the electron microscope led to the discovery that there are two basic types of living cell. One type, which is a relatively simple structure always lacking a true membrane-bound nucleus has been termed a prokaryotic cell. The second type of cell is more complex, generally larger and always has a membrane bound nucleus and is termed a eukaryotic cell. This primary division is extremely useful to us because there are no exception in that all living cells are very definitely either prokaryotic or eukaryotic. Knowledge of this tells us immediately quite a lot about the properties of the cell. Prokaryotic cells include bacteria and blue-green algae (now more usually referred to as cyanobacteria) whereas the fungi, algae, protozoa, plants and animals are all, without exception, eukaryotes. At this stage the possibility of dividing living systems into three groups or kingdoms (animals, plants and prokaryotes) was suggested. This division helped to remove the difficulty of assigning bacteria to either plant or animal kingdoms but still did not resolve the dilemma of where to place the protozoa like those mentioned earlier.

five kingdoms

The currently favoured system is the five kingdom approach. In this system the proposed kingdoms are Monera (prokaryotes), Fungi, Animals, Plantae and Protista. The major characteristics of each are shown in Table 1.1. The real difficulty of this system is to find a precise definition of the Protista.

A study of Table 1.1 indicates that the microbiologist can therefore study examples of three of the kingdoms, Monera, Fungi and Protista.

Kingdom	Primary characteristics	Example
Monera or Prokaryotae	prokaryotic cells	bacteria and blue-green algae (cyanobacteria)
Fungi	eukaryotic cells - mycelial, usually walled and septate, multinucleate	higher fungi
Animalia	eukaryotic cells - multicelluar, wall-less, aerobic, capable of complex differentiation, primarily ingestive nutrition	animals
Plantae	eukaryotic cells - multicelluar, walled, aerobic, usually differentiated, primarily photoautotrophic nutrition	plants
Protista	eukaryotic cells - ingestive or photoautotrophic	simple algae, protozoa, simple (lower) fungi

Table 1.1 The primary characteristics used to assign organisms to one of the five kingdoms.

Archaebacteria

It should be pointed our here that as our knowledge increases the grouping of organisms even at this broad level changes. As we currently learn more of a rather specialised group of bacteria called *Archaebacteria*, some researchers suggest that these should be placed separately from other living systems. Briefly they are prokaryotic in structure but have cell chemistry unlike all other living systems. Examples of this group are the halophiles which will be mentioned later in the text.

∏ Which major group have we not mentioned in Table 1.1?

viruses

The answer is the viruses. Viruses are acellular and by definition non-living. However, due to their intimate relationship with living cells and their profound effect on our lives and environment we study them within the context of biology. Because they are very small we study them within microbiology. Due to their uniqueness they have to be treated independently and their cultivation and enumeration will be studied in a separate chapter in this text.

In summary the microbiologist largely studies free-living, unicellular organisms which are virtually always too small to be seen by eye. The organisms, collectively termed micro-organisms, to be studies include all of the bacteria, all of the blue-green algae, fungi, algae and protozoa. In addition, the acellular entities called viruses are a part of microbiology.

divisions of microbiology

As we shall see in Section 1.4, micro-organisms are incredibly diverse and occur in virtually all habitats throughout the world. The diversity of the subject means that microbiologists tend to specialise in one of a number of specific areas. We can for example identify microbial geneticists or biochemists or physiologists or we can identify specialists that focus on specific groups, for example bacteriologists, mycologists (the study of fungi) or protozoologists. This text which focuses onto the cultivation of micro-organisms underpins all of these specialities.

1.3 Historical aspects of microbiology

Microbiology is a relatively young science, primarily because the small size of the organisms under study requires sophisticated microscopes or alternative technology for successful investigation. A secondary reason is that the techniques required to handle, maintain and grow micro-organisms are generally different to those required for animals and plants.

1.3.1 The influence of microscopy

Even before the occurrence of micro-organisms had been confirmed their existence had been suspected. For example around 1540 the physician Fracastoro suggested that disease was caused by 'invisible living organisms'. Around 1600, microscopes were used to observe the structures on insects and this was closely followed by the first well documented, accurate observations of micro-organisms by Antonie van Leeuwenhoek. van Leeuwenhoek was a Dutch draper who had not received any scientific training but was endowed with a great sense of natural curiosity and an expertise in constructing simple microscopes. A simple microscope is one which has a single lens as opposed to a compound microscope which has two or more lenses. van Leeuwenhoek's microscopes had to be hand held close to the eye and one looked through the lens at a sample held on a moveable mounting pin. His microscopes were capable of giving x 50 up to x 300 magnification.

van Leeuwenhoek

Π   Given that the eye can just detect structures 0.2 millimetres long and see clearly structures 1 millimetre long, what sized structures can just be seen and also be seen clearly using van Leeuwenhoek's microscope at maximum magnification? Give your answer in μm.

The maximum magnification is x 300. Thus if the unaided eye can just see structures 0.2 mm long, using the microscope structures 0.2/300 mm or (0.2/300) x 1000 μm could be seen. Similarly the microscope would enable structures (1.0/300) x 1000 μm to be seen clearly. Thus the answers are 0.667 μm and 3.3 μm respectively. The average size of a bacterium such as *E. coli* is 2 μm. van Leewenhoek would therefore be able to see bacterial cells and the somewhat larger eukaryotic cells (for example yeasts are about 6 μm in diameter). He drew his observations with great accuracy and the sketches were published regularly in the Royal Society journals in London.

There was relatively little further gain to our knowledge from microscopy for almost 200 years until around 1820 when the technology and quality of compound microscopes had improved sufficiently. Throughout the nineteenth century advances in our knowledge from microscopy paralleled advances from other means.

1.3.2 The phenomenon of spontaneous generation

Until the 1850s mankind had largely believed in the theory of spontaneous generation. Supporters of this theory believed that living organisms could develop from non-living organic matter. Various religious organisations encouraged this idea claiming that such

Redi spontaneous generation was an act of God. The Italian physician Redi proved, around 1680, in an elegant series of experiments that maggots which were thought to spontaneously generate on decaying meat were in fact only found in meat on which flies had landed and laid eggs. Meat which had been protected from files (by gauze) still decayed but no maggots were detected. van Leeuwenhoek's observation of animalcules followed this work very closely. The supporters of spontaneous generation concluded that micro-organisms did arise by spontaneous generation but that larger organisms did not. The arguments for and against continued unabated and were not finally resolved until 1861 when Louis Pasteur proved conclusively that micro-organisms do not arise by spontaneous generation. Pasteur had for some time

Pasteur been working on complex chemical conversions such as the conversion of sugars to ethanol and had proved conclusively (by a crude form of filtration) that air contained a variety of structures of the size and nature of those found in contaminated foods. It had already been established that heat kills living organisms so Pasteur set up experiments in which nutrient liquid media were heated in a specially shaped flask called a Pasteur flask (Figure 1.1). The flask was unstoppered to allow free passage of air into and out of the flask. Heat treated aliquots of media remained sterile even after prolonged periods of incubation. Pasteur correctly reasoned that any micro-organisms entering the mouth of the flask (region A) would settle and be trapped in the U bend (region B). He confirmed this by tilting one of the flasks which had been incubated for a time so that media from region C flowed along the side arm to the U bend at B. Any micro-organisms would be picked up and taken back into the region C when the flask was righted. Very quickly growth occurred. Thus organisms did not spontaneously generate in the medium, they were from aerial contamination.

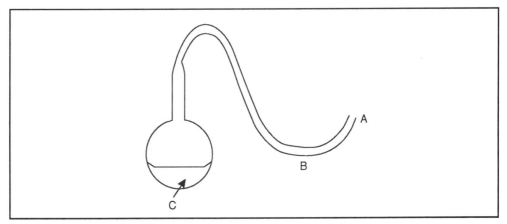

Figure 1.1 A Pasteur flask.

1.3.3 Micro-organisms as catalysts of chemical processes

Pasteur's work had also proved conclusively that the conversion of sugar to ethanol was a microbiological process. Schwann in 1837 had proposed that yeast cells carry out the process but many of the leading chemists of the time, for example Liebig and Berselius, denied the truth of this stating that chemical instability was the reason for such a change. Pasteur had been investigating an industrial process in which the production of ethanol from plant material had stopped in favour of acid production. He showed that the reason for this was the contamination by, or presence of, unwanted micro-organisms which had replaced the yeasts and which produced lactic acid. Interestingly, work by Buchner on the chemistry of the conversion of glucose to ethanol by yeasts ushered in the modern era of biochemistry when in 1897 Buchner demonstrated the production of ethanol by a cell free extract from yeast.

1.3.4 Micro-organisms as casual organisms of disease

We noted earlier that Fracastoro in 1540 suggested that diseases were caused by living organisms. However, at this time the vast majority of people did not believe this but suggested that poisonous gases, bodily malaise or even supernatural forces were responsible. Around 1825 Bassi demonstrated for the first time that a micro-organism caused disease when he showed that a fungus caused a particular disease in silkworms. He postulated that many other diseases could be due to micro-organisms and the so called 'germ theory of disease' began to gain support. Pasteur himself worked on another disease of silkworms and showed that this in fact was caused by a protozoan species.

More evidence had accumulated to support the claim that 'germs' cause disease but as yet there was nothing to prove the incidence of 'germs' causing disease in humans.

Lister and disinfection

Lister, who was a great admirer of the work of Pasteur, believed that micro-organisms did infect humans and, as a surgeon, he believed that the high incidence of gas gangrene and other post operative infections following surgery were due to entry of microbes into the wounds. He began in the 1860's to sterilise his surgical instruments by heat treatment and also sprayed phenol (a known disinfectant) on wounds, wound dressings and over the surgical area during operations. The incidence of infection fell dramatically. These pioneer experiments not only gave indirect evidence that the bacteria-killing agent phenol prevented wound infection but also initiated the early developments in disinfection generally and operating theatre protocol specifically.

Koch studies

The first direct, irrefutable evidence that bacteria cause disease was provided by Robert Koch in 1876. Koch studied anthrax, a disease of cattle which can also attack humans and is caused by the bacteria *Bacillus anthracis*. It had previously been shown by several workers that in the terminal stages of this fatal disease there are a large number of rod shaped bacteria in the blood stream of the cattle. This showed an association of the *Bacillus* sp with the infection but did not show that it was the cause of the infection.

∏   What is the next logical step to try to show that the *Bacillus* sp. causes anthrax?

The procedure that Koch used was to take some infected blood and inoculate it into an uninfected cow. The cow duly developed anthrax. Koch carried out this alternating cycle of infecting a fresh animal with infected blood and watching the symptoms of disease form. He carried this out over some twenty cycles which gives virtually conclusive proof that *B. anthracis* causes anthrax.

∏   What is the one weakness of this experiment in terms of unequivocally claiming that *B . anthracis* causes anthrax?

The answer is that although only tiny amounts of infected blood were necessary for reinfection, the samples inevitably contained blood components as well as bacteria and opponents of the 'germ theory' could claim that the infective agent was in the blood itself.

∏ What is the way forward to prove conclusively that *B . anthracis* causes anthrax?

Koch eventually managed to grow or cultivate the *B. anthracis* (we call it culturing the organism) in nutrient fluids outside the cow's body and to transfer or sub-culture the organism from flask to flask many times. Whenever he inoculated some organisms from an often transferred culture to a fresh cow the infection occurred. This then was the definite proof that *B . anthracis* caused anthrax.

Koch's postulates

From this work Koch developed a set of the minimum criteria which must be made for proving that a specific micro-organism causes a particular disease. These criteria,now known as Koch's postulates state:

• the organism must be present in infected animals but absent in healthy ones;

• the organism must be cultivated in pure culture away form the host body;

• the organism when reinoculated into the animal should initiate characteristic disease symptoms;

• the organism should be re-isolated from the diseased animal, cultured in the laboratory and should be shown to be the same as the original isolate.

Pasteur independently confirmed Koch's work on anthrax and from this time onwards the science of microbiology began in earnest.

We have so far seen the contributions made by a variety of scientists but principally by Pasteur and Koch. Pasteur has been referred to as the father of applied or practical microbiology whereas to Koch goes the honour of the title father of pure or theoretical microbiology.

1.3.5 The development of practical microbiology

The importance of the work of Koch described above cannot be over emphasised. Not only did it establish beyond question the accuracy of the germ theory of disease, it also demonstrated for the first time the use of two fundamental procedures which have since become routinely used in microbiology.

∏ Can you write down which these two procedures are?

Firstly it established the need for using laboratory culture to grow micro-organisms capable of causing infection in *in vitro* conditions. The principle of laboratory culture was also shown to be suitable for micro-organisms in general. Secondly Koch's work established the need for working with cultures of a single species of micro-organism, that is a pure culture.

The major early developments in laboratory culture techniques were generated by Koch and his group who quickly realised that liquid or broth cultures could not be used to obtain pure cultures. He observed that solid media surface would yield isolated colonies from suitably diluted liquid inocula and he reasoned that one cell should grow and eventually give rise to one visible colony. The original solidifying agent, gelatin, had two main disadvantages in that it melted at 28°C (below the growth temperature of many micro-organisms) and also it can be degraded by many micro-organisms. The wife of one of Koch's team suggested the use of agar - an agent used to solidify jelly, a compound which did not melt until heated to 98°C and which was a resistant to bacterial degradation. These two solidifying agents are still the ones in use in modern microbiology. Another of Koch's assistants, Richard Petri, in 1887, designed and gave his name to one of the basic tools of microbiology, the petri dish.

Koch used his postulates to investigate and confirm that bacteria produce many diseases, notably his work on the tubercule bacillus published in 1882. By now many pioneers of microbiology were working on a whole series of infections and by 1920 most major bacterial pathogens of human had been identified.

attenuated and vaccines

Louis Pasteur working first on cholera and anthrax and then rabies showed that cultures which were damaged or altered in some way lost their ability to infect, that is they were attenuated. Moreover in certain cases attenuated cultures when injected into healthy animals would not cause the disease but would encourage resistance to subsequent infection by a non-attenuated culture. Pasteur called the attenuated culture a vaccine (from the Latin word vacca meaning a cow). The naming was an acknowledgement of thr largely ignored work of Jenner who in 1798 had proposed vaccination of material from cowpox lesions to protect people against smallpox.

1.3.6 Nineteenth century developments in non-medical microbiology

Although the main advances in our knowledge of microbiology came from the medical field, several workers investigated the role of micro-organisms in the environment. Two workers were very prominent in this field, Winogradsky and Beijerinck.

work of Winogradsky and Beijerinck

Winogradsky pioneered our research into chemoautotrophic bacteria in the soil, that is organisms which utilise carbon dioxide as a carbon source and obtain energy by oxidation of compounds such as iron, sulphur and nitrogen compounds. Beijerinck increased our knowledge of soil microbiology enormously, pioneering work on the nitrogen fixing organisms *Rhizobium* and *Azotobacter*. Both men were responsible for the development of the enrichment culture technique; this will be considered in detail in the next chapter.

1.3.7 Virology

As mentioned earlier viruses are acellular and do not display the basic properties of living cells. In this text we have devoted a distinct chapter specifically to their culture and it is perhaps appropriate to study the historical development of our knowledge of viruses separately from cellular micro-organisms.

discovery of viruses

In 1884 Cumberland developed the porcelain bacterial filter which could successfully remove or filter out bacteria from aqueous suspensions. However, work at the turn of the century indicated that the disease causing agents of tobacco mosaic disease (in plants) and foot and mouth disease (in animals) were not bacteria because they could not be filtered out. Beijerinck called the tobacco mosaic agent a 'living germ that is soluble' and Loeffler commented that the foot and mouth causing agent belonged to the 'smallest group of organisms' as yet mentioned. The term filterable viruses (filterable life) was coined. Subsequently the term filterable was dropped and we now refer to these agents as viruses.

In 1917 d'Herelle coined the term bacteriophage for viruses which infect bacteria, a discovery made two years earlier.

In 1935 Stanley managed to crystallise the tobacco mosaic virus and showed that it was found to contain only protein and ribonucleic acid. We will explore virus structure and cultivation further in Chapter 8.

1.3.8 The development of microbiology in the twentieth century

Microbiological development occurred for most of the first half of this century in apparent isolation from the other areas of biology. Microbiologists were primarily interested in the medical aspects, characterisation of the agents of infectious disease, studies of immunity, the search for new chemotherapeutic agents and bacterial metabolic processes. Biologists were more interested in cell structure, reproduction, evolution and heredity. Some areas of potentially common ground were pursued in isolation. For example, the demonstration by Buchner in 1897 that cell free extracts prepared from yeast would convert glucose to ethanol was not connected for nearly thirty years to the almost identical process of glycolysis in muscle.

It was not until 1941 when Beadle and Tatum showed that Neurospora (a fungus) mutants could successfully be used to study genetics and biochemical pathways that microbiology began to merge with the mainstream of biology. By the 1950s the links between biology and microbiology were further strengthened when microbiologists' expertise at culturing organisms was employed for animal and plant tissue culture.

Since 1940-1950, the advantages of using micro-organisms for genetic and biochemical research has been obvious and since 1953 when Watson and Crick published their paper on the DNA double helix model, microbiology has become an essential part of what is now called molecular biology. The realisation that the relative simplicity of micro-organisms makes them easier to manipulate and alter than cells of higher organisms has allowed development of a completely new expertise, recombinant DNA technology.

SAQ 1.1

Construct a table under the headings shown below to indicate the important historical milestone in the development of microbiology. Try to put them in chronological order and attempt it initially without looking at the text. Finally check your answer with the table at the back of the book.

Date   Researcher   Nature and significance of the work

1.4 Reservoirs of micro-organisms

SAQ 1.2

Indicate whether you think that the following statements are true or false. Where appropriate try to give example of habitats or organisms.

1) Virtually every organic compound known can be degraded by one or more bacteria.

2) Micro-organisms are found everywhere on Earth from the upper atmosphere to the ocean depths.

3) Bacteria are known which will grow at 240°C.

4) Bacteria are known which will grow below pH 1.0.

5) An increase in pressure to many atmospheres (say greater than 50) has little or no effect on most bacteria.

6) Micro-organisms are absolutely essential for the natural cycling of elements such as carbon, sulphur, nitrogen and phosphorus.

7) Some micro-organisms can divide in less than 25 minutes and, given the opportunity, can produce in 44 hours from a single cell progeny equal to the weight of the Earth.

The purpose of SAQ 1.2 was to give some idea of the enormous importance and diversity of micro-organisms and their incredible powers of self-replication. If we are looking for a micro-organism to carry our a particular task for us we can be fairly confident that somewhere in the natural environment there will be an organism capable of carrying out that task. Our skill as microbiologists is to know where to look and then selection the correct methodology to isolate and store our chosen micro-organism. Although in some ways it is more satisfying to obtain the organism oneself from natural sources it may be relevant to obtain a specific strain from one of the recognised culture collections. These are fairly expensive but save on time and effort. The topic of selection of site and method of isolation or organism will be discussed in detail in Chapter 2.

1.5 Safety within the microbiology laboratory

risk of infection

In many ways this section is the most important in the whole text and should be read with great care. The intention here is to give some idea of the risks involved during work with micro-organisms; the legal recommendations and the essential requirements to be employed and finally the way to conduct oneself in the laboratory environment. All micro-organisms are intrinsically dangerous, even if they are not specific pathogens. They may be what are called opportunistic pathogens, that is they will invade by accident if in contact with open wounds etc. One of the real problems in microbiology is that for identification, teaching, research and industrial product production we have to prepare very large numbers of micro-organisms. In laboratory media it is common to reach $10^9/10^{10}$ organisms per ml. Whereas small numbers of a given species may be virtually harmless, large numbers are potentially hazardous.

We must also remember that if we have produced large numbers of micro-organisms we eventually have to ensure that they are disposed of after being rendered harmless. Sections in Chapters 2 and 9 of this text will address these issues in more detail. We must remember, however, that we have to protect not only laboratory workers but also the community at large.

The object of this section and the text as a whole is to give you an insight into *in vitro* cell culture techniques. However, the text does not under any circumstances imply or suggest that you read it and then try your own unsupervised experiments. There is no substitute for initial supervised practical work experience in properly equipped, specialised microbiology laboratories.

1.5.1 Official guidelines and regulations on safety

For many years now data have been gathered concerning laboratory acquired infections. Many thousand workers have been infected with a whole series of different infections and those instances officially notified are probably only a fraction of the total because minor infections may not be reported or even noticed. Salmonela and Brucella are the most common bacterial infections amongst laboratory workers. Fungal infections are less common than bacterial infections but those caused by viruses are increasing and now represent the most serious threat to laboratory staff. Using an example from one survey, over 20 of the microbiologists working in a series of clinical laboratories had antibodies to hepatitis B viruses indicating previous exposure to it.

It is not our intention here to review all of the national regulations concerning the use of micro-organisms, We are fortunate in that there is a large measure of agreement between countries both in terms of the objectives of the regulations and the procedures to be adopted.

This general agreement stems from the common desire to provide a safe working environment and is reflected in the recommendation arising from international organisations such as the European Federation of Biotechnology and the legal directives developed by supra-national regulatory bodies such as the EEC.

It is important at this early stage that you are aware of the principles which govern the safe use of micro-organisms. To provide you with this awareness we will briefly examine the situation within the United Kingdom.

ACDP In the United Kingdom an Advisory Committee on Dangerous Pathogens (ACDP) was set up in 1983 and their first reports was published in 1984. The second edition, published in 1990 and still entitled 'Categories of pathogens according to hazard and categories or containment' is a report which lays down essential requirements and also strong recommendations. There have been very few problems with implementing the recommendations of the 1984 booklet and they have become of model system for representing good laboratory practice. In 1989 the Control of Substances Hazardous of Health (COSHH) regulations came into force. These regulations enforce the philosophy that there should be a proper assessment of risks (of substances hazardous to health), and exposure should be adequately prevented or controlled on the basis of the results of the assessment. Pathogens are regarded as substances hazardous to health. Various European initiatives on safety are currently being developed and a wide variety of advisory pamphlets are available which provide good advice and are essential reading before practical work is undertaken.

hazard group

containment
level

The terms hazard group and containment level are now part of the day to day language of the laboratory. Hazard group refers to the degree of risk that a particular organism presents and containment level refers to the physical restraints and controls to be placed on organisms.

There are four hazard groups; 1 referring to the least dangerous organism and 4 referring to the most dangerous. The containment levels largely match the hazard group and are again based on a 1 to 4 system.

∏ What factors do you think are important when assigning a micro-organism to hazard groups 1 to 4?

The answer includes the ability or inability of the micro-organism to cause disease, the nature of the disease , the hazard to the community and whether effective treatment is required. The full definitions are shown in Table 1.3 and it should be noted that organisms fitting into hazard groups 2, 3 and 4 are referred to as pathogens. Read Table 1.2 carefully.

Hazard group 1	An organism that is most unlikely to cause human disease.
Hazard group 2	An organism that may cause human disease and which might be a hazard to laboratory workers but is unlikely to spread to the community. Laboratory exposure rarely produces infection and effective prophylaxis or effective treatment is usually available.
Hazard group 3	An organism that may cause severe human disease and presents a serious hazard to laboratory workers. It may present a risk of spread to the community but there is usually effecive prophylaxis or treatment available.
Hazard group 4	An organism that causes severe human disease and is a serious hazard to laboratory workers. It may present a high risk of spread to the community and there is usually no effective prophylaxis or treatment.

Table 1.2 Definitions of hazard groups. Note: All organisms in hazard groups 2, 3 and 4 are referred to as pathogens. Table taken from the UK Advisory Committee on Dangerous Pathogens handbook.

∏ Do you think that an inexperienced laboratory worker should be allowed to work with hazard group 4 organisms?

The answer is obviously no. It is one thing to know that an organism is hazardous, it is another to know how to use such an organism safely. Safety training can only properly be done within a laboratory context and by guidance from those experienced in the relevant techniques and procedures.

The decision as to which hazard group an organism belongs to is made by an Expert Committee and inevitably there will be controversy. Where there is doubt the Committee opt for the cautious approach. For example some workers would consider that certain hazard group 3 organisms should be in group 2. One important consideration here, however, is that because an organism is considered to be category 1 we must not relax and think that it is entirely safe. Examples of organisms in hazard groups 2, 3 and 4 have been placed in an Appendix at the back of this book. We do not expect you to memorise this list. It is there for reference. We would, however, encourage you to glance through this list as it will give you some idea as to what constitute the various hazard groups.

The regulations regarding containment are lengthy and subject to many conditions and references to special cases. However, Table 1.3 shows a summary of minimum laboratory containment requirements for the various containment levels. A study of class I to class III cabinets which are referred to in the Table will be found in Section 1.5.2.

Containment requirements	Containment levels			
	1	2	3	4
Laboratory suite: isolation	No	No	Partial	Yes
Laboratory: sealable for fumigation	No	No	Yes	Yes
Ventilation:				
inward airflow/negative pressure	Optional	Optional	Yes	Yes
through safety cabinet	No	Optional	Optional[1]	No
mechanical: direct	No	Optional	Optional	No
mechanical: independent ducting	No	No	Optional	Yes
Airlock	No	No	Optional	Yes
Airlock: with shower	No	No	No	Yes
Wash basin	Yes	Yes	Yes	Yes
Effluent treatment	No	No	No	Yes
Autoclave site:				
on site	No	No	No	No
in suite	No	Yes	Yes	No
in lab-free standing	No	No	Optimal	No
in lab-double ended	No	No	No	Yes
Microbiological (bio) safety cabinet/enclosure	No	Optional[2]	Yes	Yes
Class of cabinet/enclosure	-	Class I	Class I/III	Class III

Table 1.3 Summary of laboratory containment requirements. 1) If a Class III microbiological safety cabinet is chosen for use at Containment Level 3 it may be necessary to provide supplementary air extraction in the room in order to achieve an inward flow of air. A Class III cabinet used alone for this purpose may be insufficient. 2) Required for clinical microbiological suites.

Thus before work involving micro-organisms is undertaken we have to ascertain the risks associated with the organism. In other words we have to determine which hazard group the organism belongs. This governs the level of containment we need to operate and who can carry out the work. Before commencing work with the organism, we have to ensure that the appropriate containment facilities are available and appropriate operational procedures have been developed (see also Section 1.5.3). Although there are some differences in the details of the regulations and recommendations which are applied in different countries, standards similar to those described above are widely applied.

Overall in this section we have tried to stress the risks involved in working in a microbiology laboratory. These are obviously very variable depending on the nature of the organism an the standard of training of the operative. The guidelines given by the appropriate natural agencies are essential reading. Employers have a special responsibility for informing and training personnel in good laboratory practice within the microbiology laboratory.

1.5.2 Class I, II and II biosafety cabinets

The availability of a contained, small working environment is essential if we are using organisms of containment levels 3 and 4 (hazard groups 3 and 4) and strongly advised for group 2 organisms.

Perhaps the easiest way of containing micro-organisms is to work in a completely enclosed environment. If unlimited funds are available one can design and build a suite of rooms which are completely self contained and any workers entering pass through a double airlock system putting on sterile gowns, gloves, shoes and some form of mask before entering the working area. This dramatic type of working practice is not the norm, however, especially for the lower hazard groups.

More often we are able to operate in the smaller, bench top, containment area provided by containment (safety) cabinets.

Basically these are systems which are designed to protect the worker, mainly from aerosols which may have been released during the processing of microbial samples. Due to physical barriers or direction of air flow they prevent inhalation of organisms by the worker. Such systems must be rigorously maintained and regularly according to the manufacturers recommendations and national guidelines/regulation.

We can identify various categories of safety cabinets depending upon the degree of protection they provide. For example in the UK, three classes of cabinets. (Class I, Class II and Class III) are defined. Thus Class I and Class II cabinets should reduce the likely number of organisms which would be inhaled by at least 10^5 fold and Class III cabinets should not allow any release of organisms into the environment. Similar classification of cabinets apply elsewhere in the E.C.

Class I cabinets biosafety

Class I (Figure 1.2) cabinets are half barrier, open fronted cabinets designed to protect workers. This type of cabinet has been in regular use since the 1950's and is the most used type in the microbiology laboratory. They consist of an internally illuminated box with a glass or perspex window. Below the window is a space for the operative to insert his or her arms. Air is drawn into the cabinet via the space for putting the arms; after leaving the cabinet is passes through a double filter system. The first filter is a prefilter removing particles of 5 µm, or more and the second is a High Efficiency Particulate Air (HEPA) filter which must remove 99.97% of particles which are 0.3 µm or larger. This is obviously not perfect thus there are two consequences. Firstly the air has to be vented to the atmosphere and not back into the room and secondly, only organisms of hazard groups 1 or 2 may be used with these cabinets.

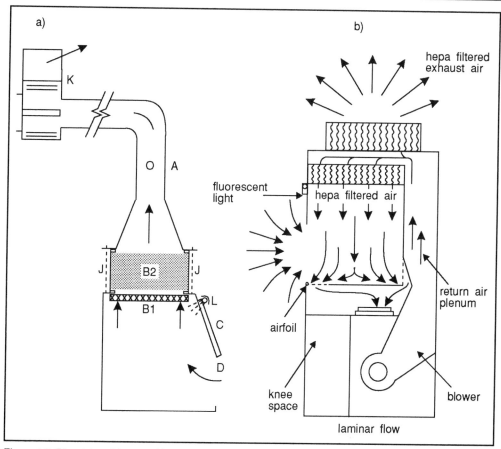

Figure 1.2 Biosafety cabinets a) Class 1 b) Class 2 (see text for a description). A exhaust airflow indicator, B1 prefilter, B2 main filter, C window, D working aperture, K extract fan, L light, JJ clamps.

Class II biosafety cabinets

Class II (Figure 1.2) cabinets are open-fronted, vertical laminar flow devices providing some protection to the worker and some to the cultures. Air is recycled via a HEPA filter within the system, some of it is continuously recycled but a portion is continuously vented to the atmosphere. The vented air is replaced by air entering as in a Class I cabinet. These cabinets are, when properly set up, suitable for hazard groups 1 and 2 organisms but not recommended for group 3 organisms and they must not be used for group 4 organisms.

The Class II vertical laminar flow cabinets should not be confused with horizontal laminar flow cabinets which have the objective of protecting the work and not the worker (for example, protecting easily contaminated animal or plant tissue culture work). Air is passed through HEPA filters at the back of the cabinet and passed horizontally over the work and out across the operator. These cabinets, as stated, are generally used for tissue culture work but they must not be used for pathogen work under any circumstances.

Class III biosafety cabinets

Class III (Figure 1.3) cabinets are gas-tight, exhaust filtered cabinets in which work is performed wearing heavy duty, gas impermeable rubber gloves. These provide maximum protection and must be regularly checked and maintained. Such cabinets are suitable for manipulating group 3 and group 4 organisms.

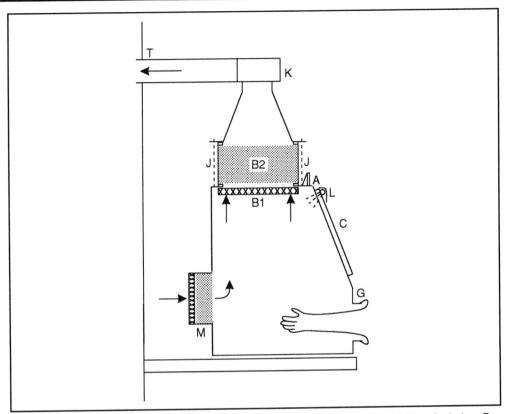

Figure 1.3 Biosafety Cabinet Class III. A exhaust airflow indicator, B1 prefilter, B2 mainfilter, C window. F front extract grille, G working port, JJ clamps, K fan chamber, L light, T extract from room, M inlet filter.

Much of the work in the microbiology laboratory involved the use of non-pathogen organisms (hazard group 1) where no special cabinets are necessary. However, we still need to utilise aseptic technique, a technique which will be discussed later in the book, and at all times employ good laboratory practice to minimise risks.

1.5.3 Local rules and operational procedures (Codes of Practice)

The national and international guidelines for the safe use of micro-organism are invariably interpreted into a set of local rules and operational procedures. These procedures will be designed to reduce the chances to infection and also reduce the extent to which materials (especially culture and reagents) become contaminated by unwanted organisms. For example each laboratory will operate a policy of disinfection and cleaning and define procedures for the disposal of used (contaminated) material. In other words, each laboratory operates under a written set of instructions (Codes of Practice) which has been formulated within the frame work of the national and international guidelines and regulations. We have hesitated to provide a specific set of examples as these are so dependent upon the particular operations and organisms used in any particular circumstances. Nevertheless before commencing any practical work using micro-organisms, you *must* make yourself familiar with, and adhere to, these local codes of practice. The point we are trying to ensure you have understood is that before you can be a practising microbiologist you should receive practical instruction within a laboratory.

1.6 Advantages and disadvantages of using micro-organisms in biotechnology

There are many practical advantages of using micro-organisms in industrial, biotechnological processes. These largely revolve around the ease and speed in which large quantities of microbes can be grown and the simplicity of these living systems allows scope for a better understanding of their metabolic processes and allows us to manipulate them in our service. We have listed the potential advantages below.

The potential advantages of using micro-organisms:

* easier to grow - more quickly, more cheaply, in unlimited quantity;

* east to manipulate in terms of their nutritional and environmental parameters;

* tend to be relatively simple at the whole cell level;

* exhibit unicellularity which makes them less demanding;

* are generally very versatile in their metabolic properties;

* allow populations of cells to be derived very quickly from a single cell giving biological heterogeneity and thus predictability;

* do not pose the ethical problems which may arise from the use of animals;

* are increasingly used in genetic engineering. For example in recombinant DNA technology in which we can alter the genome of some micro-organisms to encourage large scale production of a product desirable to us but perhaps previously not coded for by the bacterial genome.

Control of such processes has to be extremely sophisticated and some of the methods available are discussed in depth within this text. The simplicity and versatility of micro-organisms, however, means that care is required during experimental procedures. There have been notable cases of bacteria being modified to our disadvantage rather than the anticipated advantage.

Summary and objectives

This chapter has introduced the subject of microbial growth and cultivation. The book should be used as a theoretical basis for gaining laboratory experience in an approved laboratory under adequate supervision. It is not intended to be used as a do-it-yourself manual for the reader to carry out practical experiments without further practical instruction or guidance.

The chapter introduces the diversity and characteristics of the microbial world and gives a brief history of the development of microbiology concentrating particularly on nineteenth century developments. It also briefly examines the reservoirs or micro-organisms available to us, from the relatively unlimited natural environment to those held in specific culture collections. A major section dealt with the safe handling of micro-organisms in an attempt to indicate the types of regulations which must be ahered to.

Now that you have completed this chapter you should be able to:

- list the major groups of organisms which are studied in microbiology;

- discuss the historical development of microbiology, particularly the milestone of the nineteenth century and appreciate their relative importance to the subject;

- be aware of the wide variety of natural reservoirs of micro-organisms;

- describe the criteria which are applied to the four hazard groups;

- identify the appropriate class of safety cabinets to be used when conducting experiments with organisms from different hazards groups;

- be aware of the dangers of working in a microbiology;

- list some of the specific advantages and disadvantages of using micro-organisms in biotechnology.

Nutrition and cultivation of micro-organisms

Nutrition and cultivation of micro-organisms

2.1 Introduction

In Chapter 1, the introduction to this unit, the occurrence and types of micro-organisms were investigated. A brief history of the development of microbiology was then given noting some of the milestones and great microbiologists of the last 150 years. A major portion of Chapter 1 was devoted to the safe handling of micro-organisms and good laboratory practice; essentials in the potentially dangerous environment of the microbiology laboratory. Finally a summary of the relevance, advantages and disadvantages of micro-organisms as opposed to other living systems as servants of biotechnology was made.

aim of chapter   This chapter has the ultimate aim of demonstrating the strategies available for the successful isolation of any one of a variety of micro-organisms to yield a pure culture. There then follows a discussion of how to maintain and store the pure cultures.

Initially the chemical composition of microbial cells is investigated followed by the nutritional requirements of micro-organisms and the way that different culture media formulations can satisfy these demands. Several distinct categories can be identified based on these nutritional requirements and, although the terminology of these groups is somewhat complicated, it will be studied in detail because an understanding and awareness of these categories is essential to the microbiologist.

The ways in which both media formulations and the physical environment can be altered to encourage growth of relatively small groups of micro-organisms is discussed in depth. After describing the ways in which pure cultures can be isolated using conventional plating techniques, the knowledge gained in the chapter will be channelled to show examples of protocols for obtaining pure cultures.

Alternative sources of pure cultures, the commercial culture collections, will be described. Finally, after a description of the ways in which cultures can be maintained and stored, media sterilisation is discussed. Media sterilisation before use is essential for successful isolation of pure cultures and sterilisation of used culture media containing large numbers of micro-organisms must be achieved on safety grounds.

2.2 The chemical composition of the cell

The cell is the fundamental unit of living matter, it is a discrete entity which is bounded by a lipid bilayer called the cell membrane. All living cells are complicated, highly organised units in which each structure has a specific function or functions and each molecule within a given structure has a specific role. Many or all of these structures and complex molecules have to be made by the living cell, generally from simpler materials. In order to succeed, that is to be able to grow and divide, living cells must be able to obtain from their environment all of the substances which they require, not only to synthesise the molecules to produce complicated structures but also to derive the energy required for the biosynthesis. Such substances are called nutrients and must be supplied to each cell subject to its needs. Inevitably different cells have vastly different biosynthetic capabilities and therefore different nutritional requirements and it is interesting and rewarding to elucidate the specific nutrient needs of particular cells.

Before we investigate the different types of food mixtures or culture media employed it is pertinent to make some generalisations as to the elemental composition of living cells. Most living cells contain 70 to 90% by weight water, a compound which must always be available for life to be sustained.

The remaining 10 - 30% of the cell is termed the dry weight and is an unequal mixture of less than thirty of the ninety or elements which occur in the Earth's crust. It is not usual for each individual cell to contain all of the thirty relevant elements.

∏ Can you name some of the thirty elements located in living cells, particularly the six which are required in greatest quantity?

The first six elements together with their concentration in terms of percentage of the dry weight (shown in brackets) are: carbon (50), oxygen (20), nitrogen (14), hydrogen (8), phosphorus (3) and sulphur (1). Other elements include potassium, sodium, calcium, magnesium and iron together with much smaller amounts of manganese, zinc, cobalt, molybdenum, nickel, copper, chlorine and other metals and halides. The first six elements therefore constitute around 96% of the dry weight of individual cells and, *macro-elements* together with the next five in the list (potassium to iron inclusive), are often termed the macro-elements or macro-nutrients. This is because they are all required in sufficiently large amounts such that we have to ensure that they are all present in culture media.

It is obvious that cells are not merely pools of chemical elements but they consist of highly organised associations of complex, organic molecules. Living cells tend to produce a relatively small number of simple compounds called monomers which can be joined together to give a much wider variety of complex compounds called polymers.

| SAQ 2.1 | Can you fill in Table 2.1 which, when complete, will indicate the principal monomers and their resultant polymers in a typical bacterial cell? Then try to write down briefly the function of each of the polymeric groups. |

The table indicates the extensive requirements for carbon, hydrogen, oxygen, nitrogen, sulphur and phosphorus. Thus there are five more of the eleven macro-elements remaining, namely potassium, magnesium, calcium, sodium and iron.

∏ Can you suggest the roles that these five remaining macro-elements may play in living cells?

role of macro- Potassium does not have a structural role in cells but is required as an enzyme activator *elements* by several enzymes. In many cells it also plays an important role in maintaining the osmolarity of the cell. Magnesium is required to maintain the structural integrity of membranes, ribosomes and nucleic acids and it is required for the activity of many enzymes. Calcium is not essential for all micro-organisms though it has a role in stabilising cell wall structure and is essential in bacterial endospores. Like calcium, sodium is not required by all bacteria but marine micro-organisms generally require sodium for growth and halophilic organisms require very high concentrations of this

Elements contained	Monomers	Polymers	Percentage of the dry weight
carbon + hydrogen + oxygen + nitrogen + some sulphur	?	?	55%
carbon + hydrogen + oxygen + nitrogen + phosphorus	?	nucleic acids	23%
carbon + hydrogen + oxygen (+ some phosphorus)	?	lipid +phospholipid	9%
carbon + hydrogen + oxygen	?	?	5%

Table 2.1 The composition of a typical bacterial cell.

ion. Iron is probably required in smaller amounts than the other macro-nutrients, it has no structural role but is an essential component of several enzymes.

micro-elements

The remaining elements are often termed micro-elements, micro-nutrients or trace elements and are required in such small amounts that often their presence as impurities of other compounds is sufficient for the growth of cells. Remember that we must not assume that trace elements are less important to the welfare of the cell than macro-elements, they are still recognised as being essential for life. For example cobalt is found in vitamin B_{12} (itself only required in very small quantities) and zinc, copper, manganese, molybdenum and nickel are all required as enzyme cofactors or activators.

variable element requirements

The specific requirements of micro-organisms for the various elements is by no means constant. For example sodium, considered to be a macro element in higher organisms, is often required in very small quantities by bacteria. On the other hand certain organisms may need elements generally considered to be trace elements in high concentration. For example diatoms require very high concentrations of silica to construct their highly specialised cell walls.

growth factors

One final group of nutrients which we must mention are the growth factors. Growth factors are organic nutrients which cannot be synthesised by the cell and thus have to be provided in the medium. Some organisms can grow on a medium containing inorganic compounds plus glucose as carbon and energy source. However other organisms may require the addition of one or many organic compounds before growth can proceed. Note that the term growth factor is also used in higher plants and animal systems to denote 'messengers' (hormones) which stimulate growth.

∏ There are three main classes of growth factors. Can you name them?

The three classes are: amino acids - the building blocks of proteins; purines and pyrimidines - the building blocks of nucleic acids and finally vitamins - compounds required as cofactors by enzymes.

∏ If all three classes of compounds were required by an organism, which do you think would be required in the smallest quantity and why?

Amino acids, purines and pyrimidines are building blocks for biopolymer synthesis and will therefore be required in large quantities. The answer to the question therefore is vitamins as these, being cofactors for enzymes, are only required in tiny amounts.

2.3 Nutritional types of micro-organisms

autotrophs
heterotrophs

It is often useful for us to try to categorise organisms into groups, for example being able to assign organisms to nutritional categories should yield information as to the types of media required for growth. Originally organisms were divided into two categories, autotrophs which will grow in a totally inorganic environment and heterotrophs (or organotrophs) which require one or more organic compounds for growth.

With the increasing knowledge of the diversity of microbes however it became difficult to categorically define an organism as either autotroph or heterotroph. A later, more detailed classification separated organisms on the basis of both the required energy source and required principal carbon source. Thus:

- principal energy source light - phototroph;

- principal energy source chemical - chemotroph;

- principal carbon source inorganic - autotroph;

- principal carbon source organic - heterotroph.

By combining these simple pairs of alternatives it is possible to create four categories. For example an organism which uses light for energy and an inorganic carbon source is both phototroph and autotroph, hence photoautotroph.

SAQ 2.2

Using the information above can you:

1) Identify the four possible combinations (for example photoautotroph etc).

2) Define each type of organism (for example a photoautotroph is an organism which utilises light as an energy source and inorganic compounds as principal carbon source).

3) Give examples of living systems which fit into each of the four categories and estimate how extensive each category may be in terms of living systems.

It is important that you have attempted SAQ 2.2 and that you have understood the answer. Although the names are somewhat lengthy and thus daunting, an understanding of the terms helps considerably in deciding the correct medium for growing nutritionally assigned organisms.

A further complication in nutritional classification is the introduction of a third parameter along with energy and carbon sources, that is the nature of the hydrogen and electron source. We shall consider further the requirement for hydrogen and electrons in Section 2.5 when studying the effects of molecular oxygen on micro-organisms. It is suggested that the terms lithotrophic and organotrophic be used respectively to denote requirement for an inorganic or organic source of hydrogen and electrons. The nutritional classification scheme now appears to become very complicated because there are three pairs of variables indicating 2 x 2 x 2 or eight possible variations. Let us look at one example, organisms which use light energy and inorganic sources of hydrogen plus electrons and carbon are phototrophs, lithotrophs and autotroph. Collectively they would be photolithotrophic autotrophs. We might also use the term photolitho-autotrophs. Fortunately only four major possibilities exist, the ones we have already studied, together with a fifth minor group. The longer names we are now considering give more information but are not commonly used.

sources of electrons, lithotrophs, organotrophs

∏   From the information given in Table 2.2 below can you complete the two right hand columns. Use each of the words provided for column 1 and construct the relevant phrases for column 2.

Energy source	Hydrogen and electron source	Carbon source	Column 1 - shortened name	Column 2 - full equivalent name
Light	Inorganic	CO_2	a)	f)
Light	Inorganic	Organic	b)	g)
Chemical	Inorganic	CO_2	c)	h)
Chemical	Inorganic	Organic	d)	i)
Chemical	Organic	Organic	e)	j)
Required for column 1: chemoautotroph, chemoheterotroph, photoautotroph and photoheterotroph.				

Table 2.2 Requirements of each of the major nutrirtional categories of living cells.

The answers are as follows:

Column 1:
a) photoautotroph;
b) photoheterotroph;
c) chemoautotroph;
d) chemoheterotroph;
e) chemoheterotroph.

Column 2:
f) photolitho-autotroph;
g) photolitho-heterotroph;
h) chemolitho-autotroph;
i) chemolitho-heterotroph;
j) chemoorgano-heterotroph.

It is worthwhile revising these terms and making sure that you understand them.

One of the groups mentioned above has not previously been discussed in detail - the chemolithotrophic heterotrophs which are also known as mixotrophs. This group contains very few organisms, often they are chemolithotrophic autotroph which can adapt to growing on an organic source of carbon. Examples of mixotrophs are the hydrogen bacteria and *Beggiatoa*, a genus of sulphur oxidising bacteria.

mixotrophs

One final point in this section, micro-organisms in general and bacteria in particular are very versatile and adaptable and, as we have noted in this section, it is possible for organisms to grow according to more than one nutritional group. For example some algae grow photoautotrophically when subjected to a normal daily light/dark cycle, but if they are grown continuously in the dark they will grow chemoheterotrophically. Such organisms are termed facultative autotrophs. Those that will only grow in a single mode are said to be obligate auto or heterotrophs.

obligate and facultative types

2.4 The composition of culture media

We have established in Section 2.2 that micro-organisms require at the very least a series of elements and possibly some simple organic compounds in order that they may grow and divide. In addition they may require growth factors and/or organic compounds, the latter for energy and carbon source.

In Section 2.3 we began to gain an insight into both the nutritional diversity of micro-organisms and the fastidiousness of certain organisms in their nutritional requirements for successful growth.

The primary purpose of a growth medium is to enable or encourage one or more organisms to grow and divide. Optimum growth followed periodically by cell division indicates that cells are carrying out balanced growth. To exhibit balanced growth the cell has to take in and convert the simple nutrients provided into all of its required complex biopolymers: this energy-requiring biosynthetic process is termed anabolism. In many cases the same carbon source, for example glucose, may serve not only as the starting point for the synthesis of monomers but also as a source of energy. Its degradation to produce energy is carried out during catabolism. Collectively anabolism and catabolism are called metabolism. It is not within the scope of this unit to deal with metabolism in detail but we should remember that all aspects of metabolism are intimately linked to each other and the overall process is incredibly complex and generally works with near perfect precision. The first point to consider when formulating the composition of laboratory media is really - where does one begin? One may logically take the view that each element and compound required should be added to a recipe working on the premise that 'if in doubt add it' and also 'if not sure how much to add, put plenty in'. This is not a promising start, particularly when considering the relevant amounts of each component. Elements will be required at different concentrations and we should remember that some elements, particularly the trace metals may well be growth inhibitory or actually toxic at higher-than-required concentrations.

strategies for media design

In addition to establishing the relative concentration of each element care is needed in deciding in what form they should be added. Metals would not be added as filings and chlorine would not be added as a gas for example. Metals are usually added as the sulphate, phosphate or chloride, care being taken to check any problems of insolubility which may arise. Consideration of the variable oxidation state of metals may be important, for example whether to add iron as the ferric or ferrous ion. We shall discuss the basic dilemma as to whether solid or liquid media should be used in the next two sections.

From our studies to date we should have, by now, an idea of the basic components of all laboratory media - that is a mixture of simple mineral salts. It is from this starting point that we have to decide which carbon/energy/electron and hydrogen sources to add together with any specific requirements for particular organisms.

Let us now consider the information given in Table 2.3. The first column shows the mixture of simple, mineral ingredients present in each of four media labelled A, B, C and D. The four media differ from each other only in the extra ingredients added as shown in the remaining columns.

Ingredients common to all four media	Additional nutrients required for:			
	medium A	medium B	medium C	medium D
potassium dihydrogen phosphate (KH_2PO_4)	ammonium chloride (NH_4Cl)	ammonium chloride	ammonium chloride	
magnesium sulphate ($MgSO_4$)		glucose	glucose	glucose
ferrous sulphate ($FeSO_4$)			nicotinic acid	yeast extract
calcium chloride ($CaCl_2$)				
manganous chloride ($MnCl_2$)				
cuprous chloride ($CuCl_2$)				
molybdenum chloride ($MoCl_5$)				
cobalt chloride ($CoCl_2$)				
zinc chloride ($ZnCl_2$)				

Table 2.3 The composition of various culture media suitable for the growth of different nutritional types of micro-organisms.

∏   How many macro-nutrients are present in the mineral ingredients shown in Table 2.3 column 1 and what macro-nutrients are missing?

The common ingredients include a total of fourteen elements of which eight (hydrogen, oxygen, phosphorus, sulphur, potassium, magnesium, calcium and iron) are macro-nutrients and six (copper, cobalt, chlorine, manganese, molybdenum and zinc) are trace elements. The three missing macro-nutrients are carbon, nitrogen and sodium.

∏   We shall comment on carbon and nitrogen shortly but can you comment on the missing 'sodium'?

As we mentioned earlier sodium is not considered to be a macro-nutrient in many bacteria; in our example it is relegated to the level of a trace element which if required would be present in sufficient quantity as an impurity of other compounds.

SAQ 2.3	Using the data in Table 2.3:

Using the data in Table 2.3:

1) Give the reasons for adding each of the additional nutrients in each of the four media A to D.

2) What macro-nutrient is still apparently absent from medium A? How would this be supplied?

Nicotinic acid in medium C is only required in small quantities but is absolutely essential for the growth of certain organisms. For example *Escherichia coli* will grow on medium B (and of course on medium C). However its close relative *Proteus vulgaris* will grow only on medium C: it has a requirement for nicotinic acid as it cannot synthesise the compound itself.

∏   Will either organism grow on medium D?

Both *E. coli* and *P. vulgaris* will grow on medium D; thus yeast extract must contain sufficient nicotinic acid to sustain the growth of *P.vulgaris*.

SAQ 2.4

Again using the data in Table 2.3:

1) What nutritional types of organisms would you expect to grow on each of the four media assuming that molecular oxygen would be present and that light would be absent?

2) What would be the effect of having no molecular oxygen present?

defined media

undefined (complex) media

The four media A to D can be separated into two types by a further type of classification. Media A, B and C are called defined media or synthetic media in which all of the compounds contained within are identified and their precise concentrations are known. Medium D however is an undefined or complex medium in which neither the identity nor the concentration of some or all of the ingredients is known. For example we know that medium D contains yeast extract but we do not know what yeast extract contains precisely. A defined medium has advantages when investigating some aspects of metabolism; one can study for example the effect of adjusting either the concentration of a compound or exchanging one compound for another. A complex medium has the advantage that by adding a component such as yeast extract one can be more certain of providing any growth factors which may be required by a particular organism. In addition there are practical advantages. Let us consider the nutritional requirements of the bacterium *Leuconostoc mesenteroides*. This organism requires the following for growth:

• energy source (for example glucose);

• nitrogen source (for example ammonium chloride);

- six mineral salts;

- sodium acetate;

- nineteen amino acids;

- four purines or pyrimidines;

- ten vitamins.

It is obviously a daunting task to have to weigh out forty two nutrients, many in very tiny amounts. However medium D (Table 2.3) will support the growth of this organism. Thus together with glucose and ammonium chloride, yeast extract contains all the nutrients required to sustain the growth of *L. mesenteroides*.

∏   Can you think of other mixtures of materials such as yeast extract which could be used to formulate complex media?

Examples could include various peptones, for example mycological, bacteriological or meat peptones which are protein hydrolysates from fungi, bacteria and lean meat respectively prepared by partial proteolytic digestion and all are rich in simple nitrogen-containing organic compounds such as amino acids. Casein hydrolysate is a similar mixture from milk; beef extract and yeast extract are aqueous extracts containing not only amino acids but also vitamins, nucleotides and minerals. Such extracts are routinely produced commercially to very high specification for incorporation into laboratory media.

2.5 Types of culture media

In Section 2.4 we spent some time examining the ways in which defined media can be formulated to suit the demands of different organisms and how in practice it is often easier for us to use an undefined medium, particularly for the growth of nutritionally exacting micro-organisms. In this section we shall consider the various types of culture media and aim at gaining an understanding and ability to recognise their value as aids to microbial isolation of pure cultures.

pure or axenic culture   The ultimate aim for microbiologists is almost always to isolate and maintain (and to some extent identify) a pure or axenic culture. Such a culture contains only a single species of organism and may quite possibly be derived from a single cell. Pure cultures are artificial in that they rarely if ever occur in nature. Naturally mixed cultures are found; often containing many species of organisms. Thus if we obtain a sample of a mixed culture and spread it over the surface of an agar plate, each of the living cells is potentially capable of growing and producing a colony.

If we are therefore faced with the daunting task of isolating and identifying a particular type of organism we need to start with a medium which will help us, for example by encouraging the growth of the required species or by inhibiting the growth of unwanted organisms.

Examples of different types of media are:

- general purpose media;

- enrichment (enriched) media;

- selective media;

- differential media;

- diagnostic media.

SAQ 2.5   The name of each of the media above gives a clue to its purpose. Try to explain in a sentence or two what you understand to be the purpose of each medium.

Now that we have established the various types of medium available to us we can look at examples of each type of media in a little more detail.

2.5.1 General purpose media

Π Which of the media discussed in Table 2.3 could be considered as general purpose media?

Medium D is the best example of a general purpose medium. It contains a balanced array of minerals together with glucose, a commonly accepted nitrogen source and many growth factors in the yeast extract. Media B and C would support growth of many micro-organisms but not nearly as many as D due to the lack of yeast extract. Medium A is clearly not a general purpose medium due to the lack of a carbon source other than atmospheric carbon dioxide.

2.5.2 Enriched media

Medium A from Table 2.3 is an example of an enrichment medium. Inoculation of such a medium followed by incubation in the dark would allow growth of the chemoautotrophic nitrifying bacteria (organisms that use ammonia as an energy source) and incubation in the light will encourage growth of photoautotrophs.

Another example of an enrichment culture would be to use a specific but unusual carbon source. For instance we may want to isolate an organism which grows on phenol as sole source of carbon and energy. This property is largely restricted to members of the genus *Pseudomonas*. Our first task therefore would be to construct an enrichment medium suitable for supporting growth of pseudomonads and containing only phenol as sole carbon and energy source. Finally physical conditions can be manipulated, for example incubation at 55°C, irrespective of medium composition, would encourage growth of only thermophilic organisms.

2.5.3 Selective media

In a sense medium A from Table 2.3 is selective for autotrophic organisms as it does not contain an organic carbon source. Other examples which contain inhibitory substances include:

- a medium containing dyes such as crystal violet would allow growth of Gram negative bacteria but inhibit the growth of Gram positive bacteria;

- sodium azide is a metal binding agent which interferes with micro-organisms carrying out aerobic respiration. Thus azide-containing media (inhibiting aerobes) could be inoculated and incubated aerobically (thus inhibiting strict anaerobes) selecting therefore for aerotolerant and facultative types.

2.5.4 Differential media

MacConkey's agar is a selective and differential medium containing the following:

- peptone;

- lactose;

- bile salts;

- sodium chloride;

- neutral red indicator;

- crystal violet.

The medium is one alternative for the isolation and identification of enteric bacteria; that is bacteria which live in or are associated with the human gut. Bile salts and crystal violet inhibit the growth of Gram positive organisms. Enteric bacteria that can ferment lactose, for example *E. coli*, produce sufficient acid end products to decrease the pH enough to change the neutral red indicator from colourless to red. This pH drop is also sufficient to precipitate bile salts around and under the colonial growth. Non-lactose fermenters such as *Salmonella spp.* grow as small, white colonies. Thus the medium is selective for Gram negative enteric bacteria and differentiates between them on the basis of colony characteristics.

2.5.5 Diagnostic media

Many different diagnostic tests have been devised of which relatively few are in common use. Each is designed to test one or more specific property of a given cell. Such media often play little part in the primary isolation of organisms, rather they are important in distinguishing or identifying organisms at a later stage. For example several tests are available to distinguish between members of the Enterobacteriaceae; indole production one routine example would be to determine whether or not an organism produces indole from tryptophan and another example, whether acid end products or neutral end products arise during fermentation.

2.6 Culture conditions

The previous two sections have respectively demonstrated that there are many different formulations of culture media available to us. In the next section we shall study the actual isolation of pure cultures, but first we shall consider how incubation conditions can be manipulated to our benefit.

2.6.1 Solid versus liquid media

agar

gelatin

Initially we should consider briefly the choice between solid and liquid media. A solid medium often has the same ingredients as a liquid medium except for the addition of agar. Agar, first used in 1882, is a carbohydrate solidifying agent obtained from seaweed which cannot be digested by the overwhelming majority of micro-organisms. Agar dissolves in water at a concentration of around one to three per cent when heated to above 95°C. As the medium temperature cools the agar solidifies at about 45°C giving a jelly-like texture. Gelatin is the other commonly used solidifying agent but it has two major disadvantages for growth media; it liquefies above 29°C and it is hydrolysed by many micro-organisms (though the latter may be used to advantage in identification work).

∏   Make a list of the relative advantages of broth cultures and of solid (agar) media and compare your list with our discussion given below.

The major advantage of broth cultures is that virtually any volume of media may be prepared from 1 ml to industrial fermentors containing many thousand litres. Thus large numbers of micro-organisms can be obtained. The growth rate of micro-organisms is usually quicker in liquid media. Solid medium has advantages during isolation of a pure culture from a mixed culture; the characteristic appearance of colonies on agar are a useful identification aid. Finally, organisms remain viable on solid media for longer periods making storage by refrigeration much easier.

A given species of micro-organism always produces highly characteristic colonies on a particular growth medium under a specific set of growth conditions. This is advantageous to us in two ways; firstly we can tell if all isolated colonies on a plate appear to be the same, thus the same species. Secondly colony characteristics are an aid to identification. It must be noted however that training and great care are required before definite conclusions can be drawn.

2.6.2 The effect of temperature on growth

It is important to remember that it is the effect of temperature on growth only and not on survival that is being considered. Many micro-organisms will survive low temperatures at which they cannot grow.

Arrhenius
equation

The growth of micro-organisms can be considered as a series of chemical reactions and the rate at which individual chemical reactions proceed is a function of temperature obeying the relationship described in the Arrhenius equation:

$$\log_{10} v = \frac{-H^*}{2.303\,RT} + C$$

where v represents the reaction velocity; $-H^*$ is the activation energy of the reaction; R is the gas constant and T is the temperature in degrees Kelvin. A plot of velocity, v, against 1/temperature (as shown in Figure 2.1a) therefore shows a straight line of negative slope, thus a proportional relationship between increase in temperature and increase in chemical reaction. Figures 2.1b shows a similar plot of growth rate, as opposed to chemical reaction, against the reciprocal of temperature for a typical bacterium. The region 'B' of Figure 2.1b shows the proportional relationship equivalent to that in Figure 2.1a. The rapid fall in growth rate at higher temperatures (region 'A') is caused by the thermal inactivation of proteins (especially enzymes) and possibly by changes to the cell membrane. If the rate demonstrated in region 'B' was continued indefinitely as the temperature was lowered then organisms would continue to grow at temperatures well below 0°C. However the graph shows that the growth rate alters sharply and falls significantly (region 'C') resulting in most organisms ceasing growth well above the freezing point of pure water.

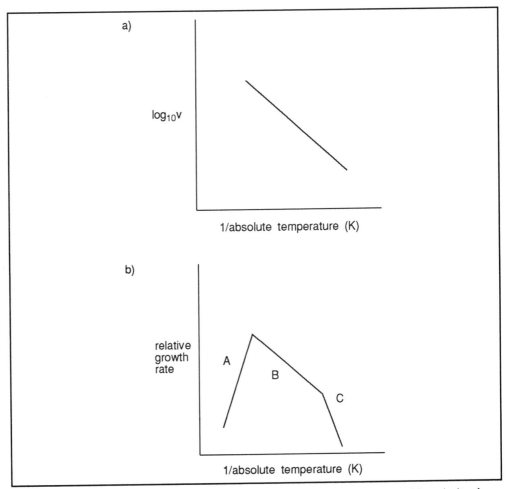

Figure 2.1 a) A typical Arrhenius plot showing the relationship between temperature and the velocity of a chemical reaction. b) A plot showing the relationship between temperature and the rate of growth of a typical bacterium.

No figures were placed on the abscissa which would indicate the temperature ranges at which living cells would grow. The overall limits from growth and division of living cells is between about 110°C and -10°C, from above the boiling point to below the freezing point of pure water! No single cell will grow over the whole range, most cells have an individual range of some 35°C though some obligate pathogens will only tolerate a few degrees of change. The examples at each end of the full range, particularly above 55°C, are bacteria where growth is restricted to very few species.

thermophiles, mesophiles and psychrophiles

Organisms are separated into three groups on the basis of their minimum, maximum and optimum temperatures, the so called cardinal temperatures. The three groups are *thermophiles, mesophiles* and *psychrophiles*. Definition of these is imprecise and there is much overlap. As guidelines however thermophiles ('heat-loving') are capable of growth without difficulty over 50 to 55°C; mesophiles have cardinal temperatures in the region of 20 to 45°C and psychrophiles ('cold-loving') grow well at 0°C. Sub-divisions are often mentioned, for example obligate psychrophiles will not grow at 20°C whereas facultative psychrophiles will.

SAQ 2.6

Can you fill in the missing gaps in Table 2.4 using examples of as many types of micro-organisms and examples of habitats as possible?

Growth temperature range			Type of organism	Examples of micro-organisms	Natural habitat
minimum	optimum	maximum			
below 0	range*	20°C	psychrophile	?	?
10°C	range*	45°C	mesophile	?	?
40°C	range*	110°C	thermophile	?	?
* the word 'range' indicates that although the optimum will be specific for a particular organism the group would have a range, for example mesophiles would have a range of 25-45°C.					

Table 2.4 Classification and occurrence of micro-organisms according to their cardinal growth temperatures.

2.6.3 The effect of pH on growth

The growth of all micro-organisms is affected by the initial pH which, in nature, may vary from pH 1.0 in acid mine waters to pH 11.0 in ammonia-rich soils. An individual organism will usually grow over a pH range of just a few units and as a generalisation fungi prefer an acid pH whereas bacteria prefer an alkaline pH. A graph of growth rate against pH typically displays a symmetrical curve as shown in Figure 2.2.

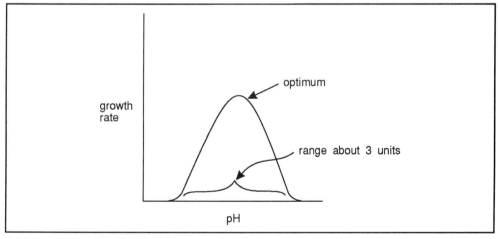

Figure 2.2 The relationship of initial pH to growth rate.

One of the problems we must consider when preparing laboratory media is not the initial pH but how to maintain a constant pH. Many organisms produce acidic end products and thus an unbuffered culture medium having an initial alkaline or neutral pH could quickly show a spectacular pH fall following microbial growth. The usual remedy is to buffer the medium, a process which is for our convenience to encourage growth of large numbers of micro-organisms but is artificial in that natural environments will generally contain unbuffered habitats.

Several buffering systems are possible and the one most often used is a form of phosphate buffer. If you recall from Table 2.3 one of the ingredients is potassium dihydrogen orthophosphate or KH_2PO_4. This is a weakly acidic salt whereas the closely related compound dipotassium hydrogen orthophosphate (K_2HPO_4) is a weakly basic salt. If acid is added to K_2HPO_4 the following reaction occurs:

$$K_2HPO_4 + HCl \rightarrow KH_2PO_4 + KCl$$

If a strong base is added:

$$KH_2PO_4 + KOH \rightarrow K_2HPO_4 + H_2O$$

The addition of phosphate to a medium thus gives a buffering capacity as well as a supply of phosphorus.

Insoluble carbonates can also be used to control pH in that they will 'take up' protons and retard development of low pHs.

$$CO_3^{2-} \underset{- H^+}{\overset{+ H^+}{\rightleftharpoons}} HCO_3^- \underset{- H^+}{\overset{+ H^+}{\rightleftharpoons}} H_2CO_3 \rightleftharpoons H_2O + CO_2$$

The usual salt used is the insoluble calcium carbonate; soluble carbonates cannot be used as they are strongly alkaline.

We should note here that some organisms can, during fermentation, produce copious amounts of organic acids and the pH changes can only be avoided by periodic or continuous addition of strong bases. This is true of some large scale fermentations in liquid culture.

2.6.4 The effect of light

The provision of light is really only important for the growth of photosynthetic micro-organisms. From a negative point of view it may also be useful to incubate cultures in the dark to avoid the growth of photo synthetic micro-organisms. In order to obtain growth of the different types of phototrophic micro-organisms, light of the correct wavelength has to be provided; eukaryotes and blue green algae absorb light at the red end of the visible spectrum whereas photosynthetic bacteria require light in the infra red region. In natural circumstances photosynthetic organisms are subjected to cyclic day and night periods but in the laboratory growth rate can be increased by providing a constant light source.

2.6.5 The effect of carbon dioxide concentration

enhanced CO_2 levels

Atmospheric carbon dioxide concentration is adequate for the needs of heterotrophic micro-organisms but autotrophs grow much more rapidly in higher concentrations of 1 to 5% carbon dioxide. This can be provided very simply in the laboratory by adding carbon dioxide to the air supply or, in the case of anaerobes providing a nitrogen:carbon dioxide mixture.

∏   What is the potentially serious problem when adding carbon dioxide to an aqueous medium?

If you are in doubt, refer back to the end of Section 2.5.3 before reading the following answer.

At the end of that section we concluded that insoluble carbonates could be added to media to remove hydrogen ions. Let us consider what happens if we add excess carbon dioxide to an aqueous medium.

$$CO_2 + H_2O \rightleftharpoons H_2CO_3$$

$$H_2CO_3 \rightleftharpoons H^+ + HCO_3^-$$

Many organisms take in their carbon as bicarbonate. Thus if bicarbonate is continuously removed a build up of H^+ ions and thus acid conditions would develop.

2.6.6 The effect of molecular oxygen concentration

In this section we have to differentiate clearly between the element oxygen and molecular oxygen, O_2. All living systems require elemental oxygen (noted in Section 2.2 to be about 20% of the dry weight of the cell) and this can be supplied in a variety of ways, for example in carbon dioxide, organic compounds, mineral salts and water.

aerobic and anaerobic respiration fermentation

The relationship of micro-organisms to molecular oxygen is however very different and allows us to distinguish between five major groups. Table 2.5 shows the five categories but before studying this table in detail we should remind ourselves of the meaning of 'aerobic respiration', 'anaerobic respiration' and 'fermentation'. Remember that it is better to understand rather than learn definitions, in the long term it is more beneficial and takes less effort.

Category	Characteristics
obligate aerobes	Organisms using molecular oxygen as a terminal electron acceptor; that is, they carry out aerobic respiration. They cannot ferment.
facultative anaerobes	Organisms which can respire aerobically or, in the absence of molecular oxygen, will grow by alternative means - generally by fermentation.
obligate anaerobes	Organisms which cannot grow in the presence of and may be poisoned by molecular oxygen.
aerotolerant organisms	Organisms which cannot use molecular oxygen but which are not directly affected by it.
micro-aerophilic organisms	Organisms which use molecular oxygen but can only tolerate the gas in low concentration.

Table 2.5 Categorisation of micro-organisms based on their relationship to molecular oxygen.

In terms of higher animals we use the word respiration to indicate breathing. Essentially we take in or inspire oxygen which is subsequently reduced and we expire carbon dioxide which is derived from reduced organic compounds. Effectively this is a form of aerobic respiration.

Aerobic respiration is the process by which chemical energy is converted from reduced organic compounds into ATP via the electron transport chain and oxidative phosphorylation apparatus; the terminal electron acceptor is always molecular oxygen. Note that ATP is not the only form of cell usable energy derived from the oxidation of substrates. We could for example cite the pH gradient established by proton pumping associated with the electron transport.

Anaerobic respiration never involves molecular oxygen. It is an energy yielding process similar to aerobic respiration in which inorganic compounds such as nitrate or sulphate serve as electron acceptors.

Fermentation is an anaerobic, energy yielding process in which organic substrates act as both electron donor and electron acceptor. There is usually a variety of end products, some more oxidised and some more reduced than the starting substrate. Again molecular oxygen is never involved.

| SAQ 2.7 | Name examples of some or all of the categories listed in Table 2.5.

At this point we should give some thought as to how we would provide the correct oxygen concentration for microbial growth. Aerobic conditions using solid media can be obtained simply by incubating surface-inoculated plates in air. Aeration of liquid media becomes increasingly difficult as the volume of medium increases . Small volumes can be adequately aerated in shaken conical flasks in which the ratio of medium volume to flask volume is kept at or below 1:5 (for example 200 ml of medium in a 1 litre flask). Oxygen is only sparingly soluble in water and as the volume is increased from around 0.5 litre to large scale industrial fermenters of many thousand litres then some form of forced aeration with sterile air or oxygen together with vigorous agitation is required.

adequate provision of O$_2$

Provision of 'virtual anaerobic conditions' which satisfies the growth of most obligate and facultative anaerobes is relatively easy but achievement of absolutely anaerobic conditions (ie the complete absence of molecular oxygen) is very difficult in practice.

Π   Can you think of ways in which a liquid medium can be maintained anaerobically?

producing anaerobic conditions

During sterilisation by autoclaving all of the dissolved gases are driven out of liquid media. Before being allowed to cool the containers are filled completely with additional sterile medium and sealed with a gas impermeable stopper. On cooling the medium is inoculated as quickly as possible and resealed.

Secondly if a gas space is required above the medium it is relatively simple to replace the air with an oxygen-free gas mixture.

Thirdly it is possible to overlay a liquid in, for instance a test tube, with another liquid which is relatively impermeable to oxygen; for example liquid paraffin.

Finally maintenance of anaerobic conditions within a fermentation vessel is accomplished by introducing an oxygen free gas such as high purity nitrogen into the culture in small but continuous amounts. However the gas may need initial treatment, for example passage through oxygen absorbers to remove the traces of molecular oxygen which are present in commercial gas preparations. Generally the apparatus would be kept under a positive pressure to avoid diffusion of atmosphere oxygen inwards.

∏ How can an anaerobic environment be provided for growth on solid media?

anaerobic jar

Generally solid medium is used in petri dishes and maintenance of anaerobic conditions by trying to seal individual dishes is not practical. Thus some sort of container is required in which plates can be placed and an anaerobic internal environment can be maintained. Nowadays for columns of up to fifteen or so plates an anaerobic jar is used. Plates are placed in the jar together with an indicator strip and a mixture of chemicals which, on addition of water, generate hydrogen and carbon dioxide. Immediately after adding water to the gas generator mixture the lid is sealed onto the jar making it gas tight. The hydrogen produced reacts with the oxygen in the jar (catalysed by an added palladium catalyst) leaving an environment of nitrogen, carbon dioxide and the excess hydrogen. A methylene blue indicator changes from blue to white in the absence of oxygen and serves as a visual marker.

anaerobic cabinet

A larger volume anaerobic environment can be provided by an anaerobic cabinet or glove box. The atmosphere within the cabinet is filled with an oxygen-free gas and the pressure maintained at slightly above atmospheric pressure. Manipulations are carried out by use of the gas-impermeable gloves fitted to the chamber wall. This apparatus can be very expensive and is obviously reliant on the quality of the gas supply being used.

2.6.7 The effect of hydrostatic pressure

barophiles

Barophilic organisms prefer to grow under high pressure - the overwhelming majority of organisms however prefer to grow at a pressure of one atmosphere. Changes in pressure, even very large increases would not affect the majority of organisms. Studies on the effect of pressure on micro-organisms have been relatively infrequent because although it is easy to generate pressure in a container, for example from a compressed gas supply, it is much more difficult to maintain that pressure and very difficult and/or expensive to maintain a pressurised environment in which micro-organisms can be manipulated.

2.7 Isolation of pure cultures

Thus far in this chapter we have established that there are several characteristic nutritional types of micro-organisms and an almost unlimited number of different culture media together with varying environmental conditions available.

If we wish to isolate a particular organism, for whatever reason, we have basically two alternative sources. Firstly, we can try to isolate one from a natural habitat or secondly we can try to purchase one from a recognised culture collection. We shall examine the second alternative in Section 2.8. If we go to a natural habitat we already know that any organism in a given sample will be in a mixed culture and we must accept that our task may be anything but easy. Pure cultures however are the basis of theoretical microbiology: they are essential to studies of microbial physiology and in relating specific diseases and infections to an individual species of micro-organism. Pure culture technology was developed around 1880 by Robert Koch and within twenty years or so pathogens of most human diseases had been isolated and characterised.

choice of medium

Our first decision has to be to choose the most suitable medium. The rationale has to centre around the perception that each organism has a preferred ecological niche. The nearer laboratory conditions are similar to that niche, the more likely it is for our chosen microbe to be the one that will grow. This is obviously a type of enrichment technique. Even the use of a general purpose medium is, to some extent, an enrichment because no single medium will support growth of all organisms. Liquid media can be used as enrichment media but we have to remember that, all other things being equal, there is equal competition for nutrients and the successful organism will be the one with the fastest growth rate.

Let us first of all look at some of the essential requirements and techniques for successful isolation of a pure culture and, having acquired this knowledge, try to devise examples of isolation programmes for specific micro-organisms.

We require:

- a knowledge of aseptic technique;

- sterile liquid and/or solid media;

- a suitable mixed culture source;

- incubation facilities.

aseptic technique

Aseptic technique is absolutely essential during manipulation of cultures and sterile media to prevent contamination by unwanted air-borne organisms. The accidental incubation of an odd cell from the air within the laboratory into a mixed culture containing many organisms at the start of our isolation may not seem too important but that contaminating organism could outgrow all others. Various videos and diagrams to explain aseptic technique are available but the only effective way to learn is to watch a demonstration in the laboratory and then try for yourselves, initially under strict supervision! Untrained personnel should never attempt isolation experiments as they can be very dangerous.

sterile media

Suitable sterile liquid or solid media are required. The word sterile indicates that the medium must be free from all living cells. Inevitably during their preparation the media will be exposed to and contain some living cells and this contamination has to be removed. We shall examine physical methods of sterilisation in Section 2.10 of this chapter. The choice of a suitable mixed starter culture and correct incubation facilities will be discussed shortly.

The usual way of isolating a pure culture is to use one of the plating methods which separate and immobilise the individual cells across a solid surface and encourage them to form discrete piles of cells called colonies. A colony of cells derived from a single cell

cloning

is called a clone. The process by which clones are produced is called cloning.

A sample can then be taken of the individual colony and the cells separated again across a fresh solid surface to check if each new colony and therefore each cell within the original colony is the same. This is because the colony from the first culture may have arisen from two or more cells which had not been well separated. Serial transfers like

sub-culturing

the one above are called sub-cultures, the technique is sub-culturing.

The production of pure cultures on solid material is invariably carried out in petri dishes; these are round, flat-bottomed dishes over which is placed a flat-topped lid of slightly larger diameter. Various sizes are available but the commonest ones are 10 cm diameter designed to hold between 10 and 20 ml of solidified medium.

2.7.1 The streak plate

A known volume of liquefied sterile, nutrient agar (above 55°C) is poured into each petri dish and allowed to cool. After it has set it is preferable to dry the agar surface to avoid subsequent movement of any motile organisms. A sample of mixed culture is introduced onto one edge of the agar plate usually with an inoculating loop and streaked across the remaining agar. One of several patterns may be employed, the idea being to progressively dilute the organisms until there are isolated individual cells. Figure 2.3 shows the streaking pattern sometimes called a quadrant streak. The area of the plate where suitable isolated colonies arise could be anywhere on the plate depending on the concentration of cells in the initial living cell preparation.

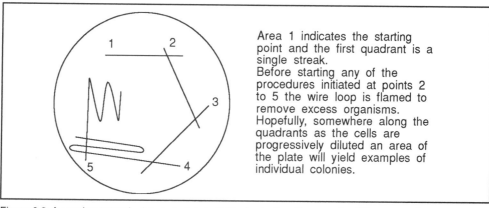

Area 1 indicates the starting point and the first quadrant is a single streak.
Before starting any of the procedures initiated at points 2 to 5 the wire loop is flamed to remove excess organisms.
Hopefully, somewhere along the quadrants as the cells are progressively diluted an area of the plate will yield examples of individual colonies.

Figure 2.3 A quadrant streak plate to isolate individual colonies. Between the different streaks (1-5) the inoculating loop is flame sterilised.

2.7.2 The spread plate

This is an alternative to a streak plate which is more often used during quantitative experiments to count the number of living cells (see the next chapter) but it can be used for isolation. In simple terms, a small volume of liquid, ideally 0.1 ml and containing around 100 living cells, is spread as evenly as possible over the whole surface of an agar plate. Following incubation individual colonies can be examined and counted or removed for sub-culturing.

2.7.3 The pour plate

A modification to the second method is used in which a small volume of original or diluted sample is placed in an empty sterile petri dish and a convenient volume of nutrient agar at about 47°C is added. The plate is swirled gently to mix the cells and agar. The agar quickly sets leaving cells on, beneath and throughout the agar depth. After incubation cells below the surface tend to produce smaller colonies than those on the surface and also growth of obligate aerobes is impeded though not totally on the bottom of the medium. This technique, like the spread plate, is often used during quantitative measurements of viable cell numbers.

For successful isolation by all methods there has to be adequate spatial separation of single cells which produce isolated colonies. Depending on the initial numbers of organisms therefore serial dilutions of the sample in sterile water may be required as a prelude to plating out.

∏ After carrying out a streak plate experiment and obtaining individual colonies, what would be the next stage or can we assume that each colony represents a pure culture?

Although we may be lucky and obtain a pure culture after a single incubation the chances are that there are still more than one cell type present. Thus a repeat plating or sub-culture of the isolated colony must be carried out as a check. Normally this process would be repeated at least three times to be reasonably certain of success.

Plating methods are suitable for the vast majority of bacteria and fungi and increasingly protozoa and algae can be grown on solid media.

2.7.4 Examples of isolation from natural habitats

Let us now consider the primary isolation of certain types of micro-organism, both in a general and in a specific sense to gain an insight into the principles used. This will be followed by examples for you to suggest an isolation policy.

The initial emphasis is almost always some sort of enrichment technique and an early decision as to whether we are seeking heterotrophs or autotrophs is usually made.

∏ What would be the most obvious effect on medium composition if we are seeking to isolate autotrophs rather than heterotrophs?

You should remember from the definitions of these two types of micro-organisms that autotrophs but not heterotrophs grow in the absence of organic nutrients. Thus our medium will not contain a carbon source, the organisms must rely on atmospheric carbon dioxide. Use of a no-carbon medium is therefore a powerful enrichment tool, it enriches for (all) autotrophs and eliminates all heterotrophs.

∏ Can you remember three other very important factors which need to be considered at the early stages of isolation?

One factor is the presence or absence of molecular oxygen. Growth under aerobic conditions enriches for obligate aerobes whilst eliminating obligate anaerobes and growth under anaerobic conditions has the opposite effect. The problem here is the large number of facultative organisms which will grow under either condition.

preferred nitrogen sources

A second factor is the nitrogen source. Heterotrophs and many chemoautotrophs prefer ammonium or organic nitrogen sources whereas phototrophs often prefer nitrate. The really selective pressure that can be used here however is a nitrogen free medium which would only support growth of nitrogen-fixing organisms.

Finally the presence or absence of light is important but not particularly useful in a positive enrichment sense. Light is essential for the growth of phototrophs and thus must be provided for their isolation. Unfortunately it is not selective in that chemotrophs are generally unaffected by light and will grow happily in its presence.

It is possible to show such an isolation scheme as a form of key. Let us look at the example in Table 2.6 of a scheme which may be used to isolate chemoheterotrophs. Our over-riding consideration must be provision of an organic nutrient which heterotrophs, by definition, require. At the second stage we make a decision to incubate in total darkness.

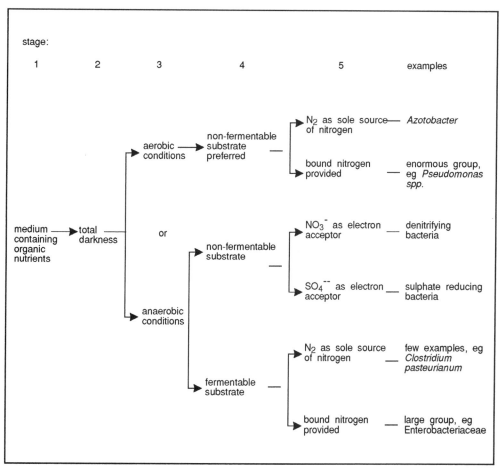

Table 2.6 Isolation scheme for the purification of chemoheterotrophs. (Adapted from Stamer, RY, Ingraham, J.L, Wheelis, M.L. and Painter, P.R. (1986) General Microbiology 5th Edition MacMillan, Basingstoke, UK)

∏ What enrichment or selection are we making at stages 1 and 2?

At stage 1 we are not really making any positive enrichment because both heterotrophs and autotrophs could grow. The medium contains organic nutrients and carbon dioxide would be available in the atmosphere. Stage 2 will effectively eliminate all obligate phototrophs due to incubation in total darkness.

Stage 3 makes the separation into aerobic or anaerobic conditions. Remember that facultative organisms will grow under either condition thus the encouragement or discouragement of such organisms will depend on the nature of the carbon nutrient(s).

Let us consider stages 4 and 5 for organisms growing under aerobic conditions. For organisms carrying out aerobic respiration either a potentially fermentable or non-fermentable substrate would suffice. A readily fermentable substrate however would encourage growth of facultative anaerobes. Thus nutrients such as glucose or other sugars should, if possible, be avoided. Exactly what to use depends on the organism being sought but could include general compounds such as ethanol or specialised substrates such as benzoate.

Looking at stages 4 and 5 for organisms growing anaerobically, the specialised groups of denitrifying and sulphate reducing bacteria can be encouraged by processing a non-fermentable carbon source under anaerobic conditions. These organisms use inorganic compounds rather than molecular oxygen as terminal electron acceptors. For denitrifying bacteria, nitrate is added as electron acceptor and ethanol or butyrate as carbon source. For sulphate reducing bacteria, elevated levels of sulphate are required together with lactate or malate as preferred carbon source.

Anaerobic incubation in the presence of a fermentable substrate encourages the growth of fermentative organisms. A small group including particularly *C. pasteurianum* are fermentative, nitrogen-fixing bacteria and can be isolated using nitrogen free medium. The large group of obligate and facultative fermentative organisms can be enriched on a medium containing a fermentable carbon source, ammonium or organic nitrogen source and anaerobic conditions. Nitrate should not be used as the medium would then encourage growth of denitrifying bacteria.

SAQ 2.8 Imagine that you want to isolate one or more micro-organisms which fix atmospheric nitrogen. Give as much information as possible indicating: your choice of medium; your choice of habitat from which to start; your choice of incubation conditions.

The answer to SAQ 2.8 indicates how much information as to incubation conditions we can deduce from the source of the material. Just one important fact to remember is that the medium suggested is an excellent enrichment medium but it is not actively selective in that non-nitrogen fixing organisms will not multiply but may well remain viable during the incubation. Thus successive sub-culture will be required, eventually on solid medium to ensure achievement of a pure culture.

Other strategies may be employed to achieve enrichment before initial incubation occurs. Consider the following information: members of the genus *Bacillus* are aerobic or facultative anaerobic bacteria characterised by the ability to produce endospores which are resistant to heat and drying. Such organisms grow heterotrophically in large number in soil.

SAQ 2.9 How would you set about trying to isolate *Bacillus spp.* paying particular attention to pre-incubation strategy?

Finally in this section we should mention that the waste products of a successfully growing organism may bring about some sort of selection process. For example

Leuconostoc species (Section 2.4) are nutritionally fastidious, fermentative organisms which require many amino acids and cofactors for growth. The use of a poorly buffered medium containing excess glucose would be a successful enrichment medium for isolation of such organisms.

∏ Why would the medium described above act as an enrichment medium for *Leuconostoc spp*?

The clues lie firstly in the fact that a poorly buffered medium is to be employed and secondly in the fact that *Leucononstoc spp.* are fermentative and are members of the group called the 'lactic acid bacteria'. These organisms will ferment the plentiful supply of glucose producing copious amounts of lactic acid which causes the pH of a poorly buffered medium to fall sharply. Non-lactic acid bacteria will stop growing as the pH drops and, at very low pH, many such organisms will be destroyed.

2.8 Culture collections

In Section 2.7 we investigated methods of isolating pure cultures from natural sources. We can accept the premise that apart from some of the industrial 'unnatural' mutants, most or all organisms are available somewhere in the natural environment. Isolation from such sources has the advantage of giving a sense of satisfaction when a required organism is successfully isolated in pure culture. However it can be laborious and unsuccessful.

use of organisms in standard test

An alternative is to purchase cultures from one of the recognised culture collections. Examples of culture collections are shown in Table 2.7. Whilst cultures are not particularly cheap to purchase there is a great saving in time and effort at the bench. In addition it may be essential in that if one wishes to conduct various tests to standard specification, for example a British Standard test, one would have to use a specific named strain from one of the culture collections. Some of the culture collections have many thousands of species in stock, others tend to be smaller and perhaps more specialised.

Abbreviation	Full name	Location
ATCC	American Type Culture Collection	Rockville, Maryland, USA
CBS	Centraalbureau voor Schimmelcultures	Baarn, Holland
CCM	The Czechoslovak Collection of Micro-organisms	Purkyne University, Brno, Czechoslovakia
CIP	Collection of the Institute Pasteur	Paris, France
CMI	Commonwealth Mycological Institute	Kew, United Kingdom
DSM	Deutsche Sammlung von Mikroorganismen	Gottingen, Germany
FAT	Faculty of Agriculture, Tokyo University	Tokyo, Japan
NCIB	National Collection of Industrial Bacteria	Aberdeen, United Kingdom
NCTC	National Collection of Type Cultures	London, United Kingdom
Many other culture collections are in existence, some small specialising in a limited type of micro-organism; some totally independent and others part of University or research institutions.		

Table 2.7 Examples of culture collections. (Adapted from Brock, T.D. and Madigan, MT 1988, Biology of micro-organisms. Prentice Hall).

2.9 Maintenance of stock cultures

maintenance by sub-culture

maintenance at reduced temperatures

In Chapters 3 and 4 of this book the growth rates of micro-organisms will be discussed in great detail. We shall see that some bacteria can double their numbers every thirty minutes indicating that in a matter of hours, enormous numbers of organisms can be produced. Thus irrespective of whether the organisms are isolated in liquid or on solid media, there quickly comes a time when the nutrients run out and waste products build up. Thus organisms become increasingly stressed in the adverse conditions and would stop growing and eventually die. To avoid losing the newly isolated cultures two possibilities arise. Firstly one could sub-culture daily to keep cultures growing and unstressed. However most laboratories maintain many hundreds of cultures and this would be totally impractical. The second alternative is to store them in some way. The usual way is to lower the temperature.

Virtually all organisms retain viability on solid rather than in liquid media at low temperature. The vast majority can be stored for days and an impressive majority of these for weeks or months on agar plates or slants in the refrigerator at temperatures of 2 to 5°C. Storage for longer times may require even lower temperatures. Providing organisms can withstand the freezing process they can be stored for very long periods in the freezer (-25°C or lower) or in a liquid nitrogen environment (-196°C). The temperature fall must be controlled very carefully and is allowed to proceed slowly in order to avoid the formation of very large ice crystals which are disruptive.

lyophilisation

Another successful technique often employed is freeze-drying or lyophilisation. In this technique frozen cells are dried by evaporating the water under a vacuum from the cultures in special ampoules. The ampoules are sealed and the powdery culture can be stored indefinitely at room temperature.

Organisms are obviously stressed by any of these techniques and require help to withstand the freezing process. For example cells to be stored in the deep freeze are usually suspended in glycerol or dimethysulphoxide for protection and cultures for lyophilisation are suspended in a milk (lactose) containing medium which helps retain viability.

When cultures are required again, frozen cultures are simply thawed and incubated; lyophilised cells are resuspended in suitable medium and incubated.

2.10 Sterilisation of culture media

sterilisation vs disinfection

The term sterilisation is a very precise one which is defined as the process which frees an object or environment of all living cells and viruses. Thus if, for example, a litre of medium contains a single living bacterium it cannot be deemed to be sterile. Disinfection on the other hand means 'the elimination of infection'. It is a less precise term because it requires a definition of infection. Successful disinfection must never be taken to indicate sterility.

physical methods of sterilisation

Chapter 9 of this book will discuss in great detail the chemical control of growth and will investigate a variety of chemical groups such as antibiotics, antiseptics and disinfectants. This section will deal almost entirely with physical methods of sterilisation, their applicability and relevance.

∏ Why do we need to sterilise fresh and used (or spent) media?

Fresh medium will obviously contain contaminants from the air and environment which have entered during the preparation. If one is involved in the primary isolation of a new organism then contaminants will, at worst, be an irritation. However once a pure culture has been obtained or bought the need for sterile media for sub-culture is paramount.

Used or spent medium must be sterilised before being disposed of. It will be potentially dangerous due to the large number of viable micro-organisms it contains.

One cannot overstress the importance of safe handling and safe removal of micro-organisms in this context. Let us consider an experiment where we were looking to isolate soil organisms. There are some extremely dangerous bacteria living in the soil, fortunately generally in low numbers. If our enrichment and isolation technique has enriched them and they are now present in large numbers, the potential risk increases dramatically. Such experiments therefore should only be conducted by trained personnel and efficient, total removal of all hazards is essential.

2.10.1 Sterilisation by heat

The most versatile, effective and commonly used physical method of sterilisation is heat. The use of heat or elevated temperature, like all physical parameters is based on the following strategy. All organisms have an optimum for growth but as one moves away from this optimum, conditions get less and less acceptable and growth first slows and then ceases. Eventually the cell will die.

Consider the graph in Figure 2.4, showing the relationship between growth rate and temperature for a typical unicellular micro-organism. The axes of the graph are unlabelled because they vary from organism to organism however the shape of the graph would be similar in all cases. There are six regions; C is the optimum growth temperature; B shows increasing growth rate with increasing temperature; D shows the steep decrease in rate due to one or more essential enzymes being inhibited by the high temperature.

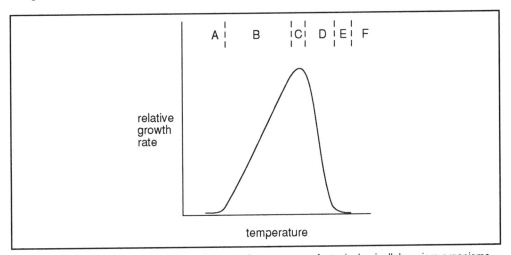

Figure 2.4 The relationship between growth rate and temperature of a typical unicellular micro-organisms.

∏   What do regions A, E and F represent?

The graph for all three regions is on the base line indicating a zero growth rate. In region A the cells are therefore not growing but, they are not killed. We could test this by raising the temperature and demonstrating that the cells would begin to grow again.

Region E is similar to region A in that cells are inhibited from growing by high temperature but are not killed. Higher temperatures still (region F) are required to kill the organisms.

biostatic   Conditions in A and E are said to be biostatic; cells are prevented from growing due to adverse conditions but if they improve, the organisms would start to grow. The temperature is so high in F that conditions are biocidal; the cells are killed. Note that low temperature or 'cold' does not kill organisms, it merely induces biostatic conditions. To sterilise, the temperature has to be high enough to reach that required at F and for long enough to kill all of the cells.

biocidal

Heat may be applied by burning - of little practical value in the laboratory except for sterilising wire loops. Secondly dry heat may be used, for example a hot air oven at 165°C for 2 hours to 3 hours. The obvious disadvantage of dry heat is that aqueous media cannot be sterilised at 165°C and, because of the high temperature needed, many medium nutrients are destroyed.

The usual sterilisation method is by wet heat.

∏ We could boil the water or solution of aqueous medium for a time - would this guarantee sterility?

The vast majority of micro-organisms and particularly those pathogenic to man are killed in boiling water. However a few viruses and vegetative cells will survive temperatures of 100°C. The real problem however is the endospores of *Bacillus* and *Clostridium* species. These will tolerate 100°C for several hours. Thus boiling water disinfects but cannot be guaranteed to sterilise.

autoclaves

Sterilisation is effected by wet heat under pressure in a piece of apparatus called an autoclave. Such a piece of apparatus is basically a metal structure varying from a simple container like a domestic pressure cooker through to sophisticated industrial autoclaves the size of rooms. Steam is generated internally or piped in from an external source and the air inside the autoclave is displaced. By closing the exit valve the pressure is allowed to rise to two atmospheres at which point water boils at 121.6°C. This temperature, if maintained for 15 minutes, is high enough to sterilise containers of medium within the autoclave. The steam generator is then turned off, the autoclave allowed to cool and, after the pressure has returned to its ambient level the autoclave may be opened.

sterilisation of glucose

Three possible pitfalls need to be avoided here. Firstly all of the air must be replaced by pure steam in the autoclave interior to make sure that the temperature rises to 121.6°C. Secondly large volumes of media have to be pre-warmed, made up using hot water or autoclaved for more than 15 minutes to ensure that all of the medium reaches the correct temperature. Finally there is the problem of the adverse effect of heat on some chemicals. For example glucose, a common medium constituent, tends to char if sterilised in concentrated solution at 121.6°C. This has deleterious effects on bacteria and should be avoided. Thus the nutrient has to be sterilised as a dilute solution or by alternative means. It is common to sterilise glucose separately as it may be partially decomposed in the presence of other ingredients, particularly phosphates.

2.10.2 Sterilisation by filtration

sterilisation of heat labile nutrients

Filtration is particularly useful for the sterilisation of aqueous solutions of heat labile nutrients. A variety of filters are available but the one most commonly used for medium sterilisation is the membrane filter. Solutions are passed through a special filter into a previously sterilised flask and are thus themselves sterilised. Filters are very specialised discs made of cellulose acetate or cellulose nitrate which contain large numbers of holes or pores. Commercially, filters of a wide range of pore size can be purchased however a given filter has a fairly uniform pore size. Bacteria may be up to 10 μm in diameter and down to 0.3 μm. The real problems here are virus particles which may be as small as 10 nanometres (0.01 μm) and filtration cannot guarantee virus removal. Filters with pore sizes of 0.2μm are commonly used to remove metabolically active cells.

2.10.3 Sterilisation by radiation

Radiation will be discussed in some detail in Chapter 5 of this book however it is relevant to mention at this point that the damaging effects of some forms of radiation can be channelled to sterilise environments. Radiation is the emission and propagation of energy in wave form through a substance or space. Emission in the visible region and at longer wavelengths (infra red, radar and radio waves) are not harmful.

UV light

However wavelengths just shorter than the visible range, known as ultra violet light, dramatically damage the nucleic acids of living cells. Of particular effect are wavelengths between 250 and 270 nm which cause production of lethal mutations within cells thus killing them. Ultraviolet light, though very effective, has limited practical value as a sterilising agent because it is absorbed by water, glass and most solid materials. A further disadvantage of strong ultraviolet light is its danger to humans potentially causing skin cancers and eye damage.

Gamma radiation

Gamma rays, of shorter wavelength than ultraviolet light, are even more dangerous to humans but when controlled are very useful to us because they are more penetrative than ultraviolet rays. Gamma rays are usually generated from a cobalt 60 source and are used for sterilising media, some pharmaceuticals and medical equipment such as prepacked syringes. Chemicals are ionised and form free radicals of which *OH, the hydroxyl radical, is the most important one formed. These radicals react with and damage both proteins and nucleic acids.

2.10.4 Sterilisation by chemicals

ethylene oxide and its disadvantage

Some chemicals are available for sterilising media and apparatus in situations when heat or radiation may not be appropriate. For example petri dishes can be sterilised using ethylene oxide. This is a chemical which is toxic to micro-organisms and volatile with a boiling point of 10.7°C. It is applied as a liquid at 4°C in an apparatus similar to an autoclave. After an appointed time the temperature is raised and the ethylene oxide gas displaced by sterile air. Ethylene oxide however is unstable and in aqueous solution decomposes to ethylene glycol which may be undesirable in the culture media. Ethylene oxide is also explosive and toxic to humans; thus its disadvantages are significant and render it a relatively unpopular option.

After media have been used, they will generally contain large numbers of micro-organisms which must be killed before any disposal is attempted. Because media are never reused, the easiest and cheapest guaranteed method of sterilisation is routinely used and this is steam sterilisation in the autoclave.

Summary and objectives

This chapter has explored the relationship between the chemical composition and nutritional requirements of living cells on the one hand to the formation of laboratory culture media on the other. The ways in which physical factors can be manipulated to make media selective for certain micro-organisms have been discussed. Examples of the strategies available for the isolation of pure cultures from a variety of sources were developed to a successful conclusion. Finally the maintenance, storage and safe disposal of micro-organisms were considered.

Now that you have completed this chapter you should be able to:

• list the major elements and compounds found in living cells;

• name and recognise the groups of micro-organisms based on their nutritional requirements;

• explain the differences between general purpose, enrichment, selective, differential and diagnostic media;

• describe the ways in which environmental conditions may be manipulated in the quest for pure cultures;

• derive a protocol for isolation of different types of pure culture;

• use the resources of culture collections if isolation of pure cultures from natural sources is not applicable;

• describe the efficiency of the various methods available for sterilising fresh and spent media;

• appreciate the need for media sterilisation on the grounds of safety.

The estimation of biomass

The estimation of biomass

3.1 Introduction

Biomass is a general term which refers to biological material in culture. Measurement of the amount of biomass is usually necessary to quantify physiological studies in biology. For example, an increase in biomass concentration with time is often an indicator of growth. The most commonly used methods of biomass estimation for

cell numbers

cell mass

growth studies are measurement of cell numbers and measurement of cell mass. These techniques may not be appropriate for some studies and methods based on determination of specific cell components or rate of substrate consumption are often used. For instance, to measure the microbiological quality of water its ATP content can be measured, with the assumption that ATP only occurs in living organisms. The amount of ATP then serves as a measure for the number of living cells or their activity.

We must remember however that an increase in biomass may not be accompanied by an increase in cell numbers. For example many bacteria produce capsules which are usually many times the weight and size of the original cell. This increase in weight and size however would not be accompanied by an increase in cell numbers and would not indicate growth.

In this chapter we will consider different approaches to the estimation of biomass and the factors influencing the choice of approach. All the approaches considered in this chapter have been used for measuring growth of microbial populations. You will see that the choice of approach for a particular study is a crucial decision to make and often the limitations of methods make the decision difficult.

For some approaches many different methods can be used. However, it is not the intention in this chapter to provide exhaustive details of all these methods.

3.2 General problems of accuracy versus precision

accuracy and precision

As with all measurements the estimation of biomass is subject to general problems of accuracy and precision. These two terms are distinct but are defined in related ways. Accuracy refers to the difference between the value of a set of measurements and the true or actual value of the parameter under investigation. On the other hand, precision refers to the magnitude of the difference between the repeated measurements themselves without reference to the 'true' value.

∏ For each of the targets shown, describe (good/poor) the accuracy and the precision of the bullet holes.

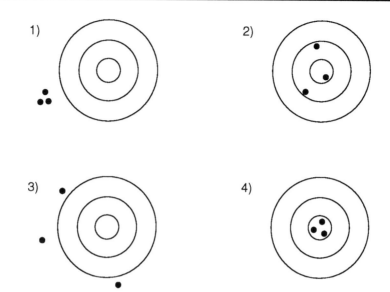

You should have concluded the following:

1) Poor accuracy, good precision. 2) Good accuracy, poor precision. 3) Poor accuracy, poor precision. 4) Good accuracy, good precision.

In the estimation of biomass the problems associated with accuracy of measurement is particularly important. This is because of differences in the definition of growth (an increase in biomass or increases in cell number) and because major errors are known or suspected in all the methods available to us. Therefore, it is usual to compare results from different methods with one that is considered to be the most reliable. However, the choice of most reliable method is often made on rather subjective grounds which includes estimates of precision and reproducibility in the hands of different investigators.

As you work through this chapter you should remember that validation of the accuracy of any measurement is not possible in an absolute sense.

3.3 Measurement of cell numbers

3.3.1 Direct counting

direct
microscopic
counting

The most obvious way to determine cell numbers is by direct microscopic counting. The small size of most microorganisms and their large population densities make it necessary to use special chambers such as Petroff-Hauser or haemocytometer chambers to count the number of cells in a sample. The former chambers are designed for counting smaller, generally bacterial, cells whereas the latter are more suitable for counting the larger cells of eukaryotic micro-organisms and, as the name implies, blood cells. These specially designed slides have chambers of known depth with an etched grid on the chamber bottom (Figure 3.1).

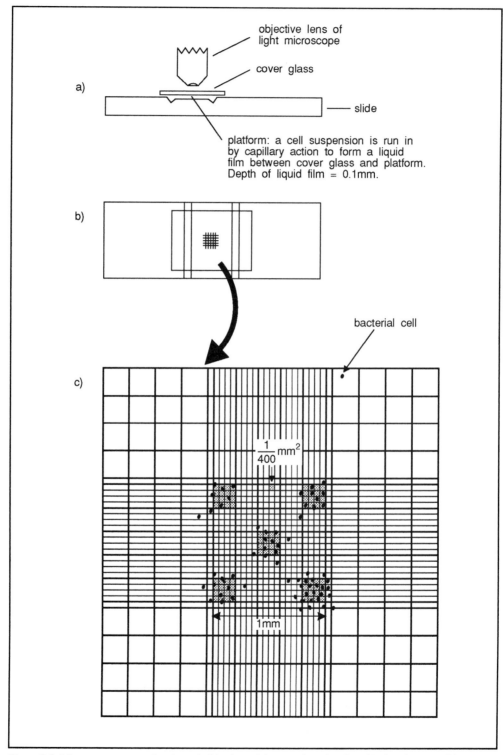

Figure 3.1 A haemocytometer chamber. a) side view of the counting chamber, b) plan view, c) grid magnified. Note we have not put cells evenly over the grid; this is to make an intext activity simpler. Normally, of course, we aim to have cells evenly distributed across the whole grid.

The number of cells in a sample can be calculated by taking into account the chamber's volume and any sample dilution required. In practice, a reasonable count is obtained as follows:

- the four large corner squares and the centre square are counted (total of eighty small squares). This minimises errors due to possible uneven distribution of cells across the grid;

- if necessary, samples are diluted so that the squares contain a readily countable number of cells. However, in order to minimise statistical errors, due to counting small numbers, each of the large squares must contain at least five cells. The more cells we count, statistically the more accurate our result should be;

- cells in contact with the boundary of each of the large squares are counted for two of the sides and ignored for the other two sides.

Π Examine Figure 3.1 and determine the number of cells ml^{-1} for a yeast culture which was diluted five-fold prior to counting. You should note that in practice yeast cells would be distributed evenly across the grid. The data you need is on Figure 3.1.

Counts per large square (north and west sides counted, south and east sides ignored - note we count any cells on the top and left hand edge of the square):

Top left = 6

Top right = 7

Centre = 7

Bottom left = 6

Bottom right = 11

Total = 37 cells in eighty small squares.

Volume associated with a small square = area x height = 1/400 x 0.1 = 0.00025mm^3.

There are 1000mm^3 in 1 cm^3 (ml), so a small square volume = 0.00025/1000 = 0.25 x 10^{-6} ml.

Number of cell in suspension counted = (37/80) x (1/0.25 x 10^{-6}) = 1.85 x 10^6 cells ml^{-1}.

Because the culture was diluted five fold prior to counting then:

Number of cells in yeast culture = 1.85 x 10^6 x 5 = 9.3 x 10^6 cells ml^{-1}.

You should note that the count is expressed to only one decimal place to reflect the accuracy of the method.

There are three main limitations to direct microscopic counting:

- it is tedious and therefore not suitable for large numbers of samples;

- it is not very sensitive;

- living cells cannot be distinguished from dead cells.

∏ Use the information presented in Figure 3.1 and the discussion given above to determine the minimum number of cells ml^{-1} required to obtain a reasonable count of a sample.

For a reasonable count a minimum of five cells per large square is required. This is obtained from a cell suspension containing 1.25×10^6 cells ml^{-1}.

viability staining

Direct microscopic counting, by the method described, gives a total cell count since it does not distinguish between viable (living) and dead cells. However, if direct microscopic counting is combined with viability staining it is often possible to determine the proportion of viable cells in a population. There are several types of viable stains which work in different ways. For example, methylene blue is reduced to a colourless form by cells capable of respiring, dead or non-respiring cells will stain blue. For yeast cell suspensions eosin is often used. Viable yeast cells are impermeable to this dye whereas dead cells stain red.

electronic counters

Larger unicellular micro-organisms such as protozoa, algae and yeasts can be counted directly with electronic counters, such as the Coulter counter. Here, the cell suspension is forced through a small hole or orifice which has an electric current flowing across it. Every time a cell passes through the orifice electrical resistance increases (or conductivity drops) and the cell is counted. Obviously to obtain an accurate count the orifice must be small enough to ensure that only one cell passes through at a time. The main advantages of the Coulter counter are that it is not tedious and gives accurate results within a few minutes for larger cells. The instrument has the added benefit of being able to estimate the size of individual cells from the change and duration of the drop in conductivity. However, a limitation of the Coulter counter is that it cannot often be used to count bacteria.

∏ Can you think of a reason why the Coulter counter is not ideally suited to counting bacteria?

For bacterial counting problems arise due to the requirement for a small orifice. Small non-biological particles in the samples, cell clumping or chains of bacterial cells can lead to inaccurate cell counts or even blocking of the orifice.

mycelial fungi

One group of organisms which cannot be estimated by numerical methods is the mycelial fungi. These are organisms which grow not by cell division but by elongation and branching of filaments. Filaments and branches are very variable in dimension and it would be impossible and meaningless to try to estimate and quote numbers in this case. We can however obtain a rough idea of the extent of growth by measuring colony diameter - a very crude, simple technique. Some of the alternative techniques listed in Sections 3.4 to 3.6 will still be applicable.

colony diameter

3.3.2 Viable counting

viable counting

In many studies we are interested in counting only living cells, that is, those able to grow and reproduce. The enumeration of micro-organisms by viable counting is based on the assumption that each viable micro-organism in a suspension will give rise to a single colony after incubation on a suitable medium under favourable conditions. After incubation the number of colonies formed is counted and used to estimate the number of viable cells in the original suspension.

The viable count is considered to be a minimum count because the number of colonies on the plate represents only those micro-organisms that can multiply under the conditions that have been established. For example, a soil sample may contain both aerobic and anaerobic bacteria, but if the plates are incubated in the presence of air, then only aerobic and facultatively anaerobic bacteria will grow. In addition, not all cells give rise to a colony because certain micro-organisms have a tendency to clump or aggregate. When plated onto a suitable culture medium a clump will give rise to only one colony, regardless of how many cells there are in the clump. For this reason microbiologists express results obtained by viable counting as colony forming units (CFUs) ml⁻¹ rather than cells ml⁻¹.

colony forming units

Ideally, only plates containing 30 to 300 colonies are counted because this will enhance the accuracy of the count. If the sample contains too many micro-organisms (many more than 300 per plate) not only will the likelihood of a colony being formed from a clump of cells increase but the colonies will be too crowded and will cover the entire plate. For this reason the samples are usually diluted. If the sample is too dilute (only a few cells per plate), however, the result will not be statistically valid.

spread-plate

pour-plate

Viable counts are usually carried out by spread-plate or pour-plate technique. In the spread-plate technique, a small sample (usually 0.1 ml) is aseptically spread over the surface of an agar plate containing an appropriate medium. In the pour plate technique the sample is either mixed with melted agar and the mixture poured into a sterile plate (Figure 3.2) or pipetted into an empty sterile plate, the agar added and the plate gently swirled. The former method has the advantage that it may give better distribution of colonies within the agar whereas the second method (described more fully in Chapter 2 of this unit) puts less stress on heat sensitive organisms. The organisms are thus fixed within the agar gel and form colonies. With pour plates, larger sample volumes can be used, 1.0 ml or more, and heavy slurries or suspensions can be used. A potential source of error with pour-plates is that organisms affected (inactivated) by the brief heating in the melted agar (around 48°C) will not be counted. However, viability of most bacteria and fungi is not affected at this temperature.

∏ Make a list of possible sources of error in viable counting by the pour-plate technique.

Possible sources of error include:

- pipetting errors;
- cell clumping;
- loss of viability due to the temperature of the melted agar;
- medium unable to support growth of micro-organisms;
- loss of viability in diluent;
- inadequate mixing of dilutions (this is very important, remember cells can quickly settle out of suspensions);
- carry-over of cells from low dilution to high dilution;

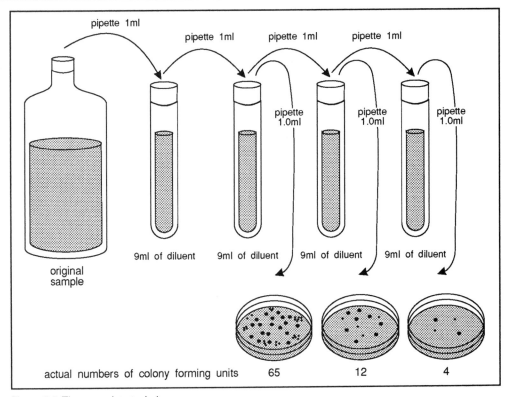

Figure 3.2 The pour-plate technique.

- heavy contamination of higher dilutions or of agar medium;

- possible antagonism between different species of micro-organisms.

Care must be taken to minimise error from these sources. For example, the carry-over of cells from low to high dilution is minimised by using a fresh pipette for each transfer when creating the dilution series. CFUs ml^{-1}, estimated by pour-plate and spread-plate techniques, is given to only one decimal place to reflect the potentially large error associated with the measurement.

Π   From the information presented in Figure 3.2, determine the number of CFUs ml^{-1} in the original sample.

We can see from Figure 3.2 that 65 colonies formed at a 100-fold dilution. Since this is within the desired 30 to 300 colony per plate range, this dilution is selected for the count. The original sample therefore contains 65×10^2 or 6.5×10^3 CFUs ml^{-1}. It should be noted that in practice, the measurements are done in duplicate.

You should note that both large and small colonies on the plate are counted. The difference in colony size in the pour-plate technique can be accounted for by most cells being fully entrapped within the agar gel (small colonies) but some cells being able to develop on the surface of the agar (large colonies). The count is expressed to only one decimal place to reflect the accuracy of the count.

<table>
<tr><td>

SAQ 3.1

</td><td>

The number of colony forming units in a growing bacterial culture was estimated by the spread plate technique using 0.1 ml volumes of suitable dilutions onto nutrient agar plates. At the beginning of the experiment, the number of colonies formed on three replicate plates was 126, 93 and 81 at 10^6 dilution. After the bacterial culture was incubated for 8 hours, a further sample was taken for viable cell counting. In this case, the number of colonies formed on three replicate plates was 102, 91 and 107 at 10^8 dilution. Calculate the number of CFUs ml^{-1} in the original culture before and after incubation.

</td></tr>
</table>

A major limitation of viable counting by the spread-plate and pour-plate technique is the need for a long incubation time; plates are usually incubated for 1 to 2 days before colonies are counted. However, the number of viable cells in a population can be

slide culture   determined within only a few hours by a technique known as slide culture, which combines microscopic examination and growth. A sample of the population, appropriately diluted, is spread over a thin layer of agar medium on the surface of a sterile glass slide. The inoculated slide is then incubated for a period sufficient for several cell divisions and is examined microscopically. Under these conditions, the viable cells develop into microcolonies and are readily counted.

In comparison with direct counting, the viable counting methods are far more sensitive. For example, a suspension containing 1 cell ml^{-1} would be detected by the pour-plate technique. For samples containing less than 1 cell ml^{-1} it is possible to perform a viable

most-probable-   count determination by the most-probable-number (MPN) technique. This technique
number   uses only liquid culture for growth and therefore can be used to count bacteria which will not grow on any agar medium, requiring liquid culture for growth. At these low levels of cells in a sample, results are expressed in terms of probability of finding a cell rather than in terms of the cell concentration. For example, if there are 3 cells in 10 ml of liquid sample the concentration could be expressed as 0.3 cells ml^{-1}. However, because cells are individuals we say that there is a probability of 0.3 for each 1 ml portion of the sample containing a viable cell. If the sample was divided in to 1 ml volumes, three would be expected to contain one viable cell and seven would be expected to contain no cells. This is the basis of the MPN technique.

In the MPN technique replicate portions of different volumes (eg 0.1, 1.0 and 10 ml) are inoculated into separate tubes containing fresh growth medium. After incubation, tubes which received a viable cell will show growth (cloudiness or turbidity) and those which did not will not show growth. The number of positive tubes (those showing growth) in each replicate can then be used to determine the MPN of cells in the sample by means of MPN tables. These are statistical tables constructed for use with three, five or ten replicate growth tubes. Three replicate tube MPN tables are shown in Table 3.1. Greater reliability is obtained by using large numbers of replicate tubes but at the expense of more time and materials in conducting the measurement.

Tubes positive			MPN	Tubes positive			MPN	Tubes positive			MPN
10 ml	1.0 ml	0.1 ml		10 ml	1.0 ml	0.1 ml		10 ml	1.0 ml	0.1 ml	
0	0	1	3	1	2	0	11	2	3	3	53
0	0	2	6	1	2	1	15	3	0	0	23
0	0	3	9	1	2	2	20	3	0	1	39
0	1	0	3	1	2	3	24	3	0	2	64
0	1	1	6	1	3	0	16	3	0	3	95
0	1	2	9	1	3	1	20	3	1	0	43
0	1	3	12	1	3	2	24	3	1	1	75
0	2	0	6	1	3	3	29	3	1	2	120
0	2	1	9	2	0	0	9	3	1	3	160
0	2	2	12	2	0	2	14	3	2	0	93
0	3	0	16	2	0	2	20	3	2	1	150
0	3	0	9	2	0	3	26	3	2	2	210
0	3	1	13	2	1	0	15	3	2	3	290
0	3	2	16	2	1	1	20	3	3	0	240
0	3	3	19	2	1	2	27	3	3	2	460
1	0	0	4	2	1	3	34	3	3	2	1100
1	0	1	7	2	2	0	21	3	3	3	1100+
1	0	2	11	2	2	1	28				
1	0	3	15	2	2	2	35				
1	1	0	7	2	2	3	42				
1	1	1	11	2	3	0	29				
1	1	2	15	2	3	1	36				
1	1	3	19	2	3	2	44				

Table 3.1 Most-probable-number tables (usually referred to as McCradie's Tables in acknowledgement of their devisor). MPN 100ml^{-1}(the number of cells per 100 ml), using three replicate tubes, each set inoculated with 10, 1.0 or 0.1ml of sample. Thus if one of the 10 ml tubes, one of the 1 ml tubes and one of the 0.1 ml tubes each showed growth the most probable number of organisms in 100 ml of the sample would be 11. (see text for details)

∏ Use Table 3.1 to determine the MPN per 100ml^{-1} in a sample which gave the following results.

Volume of inoculum (ml)	Growth (+); No growth (-)
10	- + -
1	+ + -
0.1	- + +

The number of tubes positives in the 10, 1 and 0.1 ml replicate sets of three are 1, 2 and 2 respectively. We can see from Table 3.1 that this combination of positives gives a MPN 100ml^{-1} value of 20.

SAQ 3.2

For each of the descriptions, labelled 1) to 10), select appropriate counting methods from the list provided below.

1) Most sensitive method.

2) Least sensitive method.

3) Quickest method.

4) Not suitable if count required within a few hours.

5) Estimates viable cells only.

6) Can be used to estimate proportion of viable cells in a sample.

7) Not suitable for the enumeration of heat labile (sensitive) cells.

8) Not suitable for samples containing non-cellular particles.

9) Count influenced by cell clumping.

10) A relatively large sample volume is used.

Method

Pour plate; Spread plate; Coulter counter; Haemocytometer; MPN; Slide culture.

3.4 Measurement of cell mass

Population growth is accompanied by an increase in the total cell mass as well as in cell number. Techniques for measuring changes in cell mass can therefore be used to follow growth. A limitation of this approach is that the techniques do not distinguish between the contribution of live and dead cells to the total mass. Cell mass measurement can be performed by direct or indirect techniques.

3.4.1 Direct procedures (weighing)

The only direct ways to measure cell mass is to determine the dry weight or the wet weight of cell material.

dry weight   Dry weight measurement involves three stages;

• separating the organism from the medium;

• washing the cells;

• drying the biomass.

Organisms may be separated from the medium by filtration or by centrifugation. Washing the biomass should then be carried out in such a way as to prevent lysis of the organism through rupture or osmotic shock. This may occur if the biomass is washed with water, especially when the organisms are taken from a rapidly growing culture. The precaution against lysis is to wash with a near isotonic saline and correct for the dry weight of the salt present after drying. Usually biomass is dried at 80°C for 24 hours or at 110°C for 8 hours. With modern equipment, such as a microwave oven for drying, it is possible to determine the dry weight of the biomass within thirty minutes.

The main limitation of a dry weight measurement is that it is relatively insensitive and inaccurate. With routine laboratory equipment it is difficult to weigh with accuracy less than 1 mg, yet this dry weight may represent as many as 5 billion (5×10^9) bacteria. To get a reasonably accurate result we usually need to use about 50mg of cells or more.

∏   List circumstances where sample dry weight measurement is not suitable for the estimation of biomass.

The sort of circumstances we hoped you would think of are:

• small sample size;

• medium contains an indeterminate amount of other solids besides the biomass;

• result required in a matter of minutes.

wet weight   Wet weight determination also involves the separation of cells from the medium and washing. The weight of the wet biomass is then determined directly. The wet weight will include both intracellular and extracellular water. In the bacterium *Escherichia coli*, for example, the extracellular water volume of close-packed cells has been estimated to be 10.2% of total volume and water represents 75% of the total cell weight.

For wet weight measurements to be of any use the centrifugation or filtration method used to pack down the biomass must be carefully standardised, since this influences the extracellular water content. The wet weight method is not as accurate as the dry mass method but is quicker.

3.4.2 Indirect methods

spectro-
photometers

More rapid and sensitive techniques of determining cell mass are based upon the fact that microbial cells scatter light striking them. Because microbial cells in a population are of a roughly constant size, the amount of scattering is proportional to the concentration of cells present, within certain limits. When the concentration of bacteria reaches about 10 million (10^7) cells ml^{-1}, the medium appears slightly cloudy or turbid. Further increases in concentration results in greater turbidity and less light is transmitted through the medium. The extent of light scattering can be measured using a spectrophotometer and is almost linearly related to bacterial concentration at low absorbance levels. Absorbance (A) is defined as the logarithm of the ratio of intensity of light striking the suspension (Io) to that transmitted by the suspension (I).

$$A = \log \frac{Io}{I}$$

Light scattering by a bacterial suspension is illustrated in Figure 3.3.

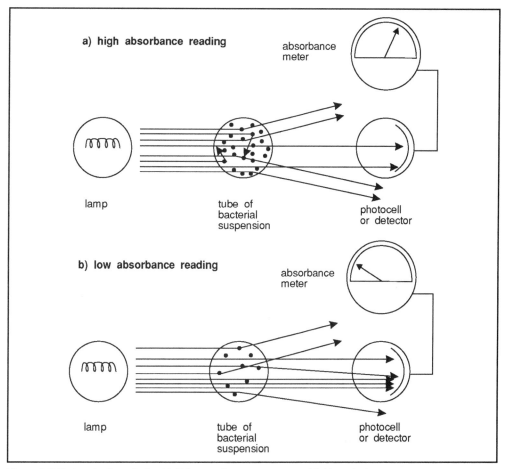

Figure 3.3 Determination of microbial biomass by measurement of light scattering.

A spectrophotometer is convenient for estimating culture density. It can also be used as a rapid and accurate method of measuring the dry weight of bacteria per unit volume of culture. This of course can only be achieved using a standard curve, such as that shown in Figure 3.4. The values shown in this figure are fairly typical. Note that we require about 0.1mg dry weight of cells ml^{-1} to get a reasonable absorbance. Thus if we use a 1 ml cuvette in the spectrophotometer, we need to use about 0.1mg dry weight of cells. Thus gives some idea of the sensitivity of the method.

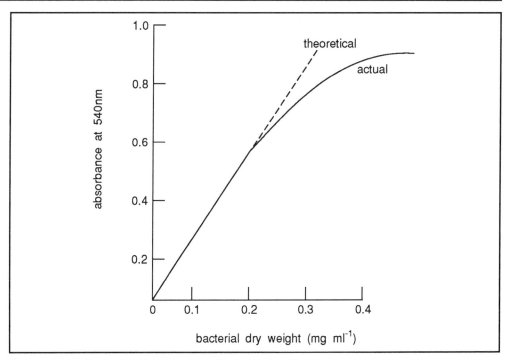

Figure 3.4 Absorbance of light as a function of bacterial dry weight.

⫫ We can see from Figure 3.4 that the relationship between absorbance and dry weight becomes non-linear at relatively high dry weight values. Can you think of a reason for this?

As the number of cells in a suspension increases a greater proportion of cells do not contribute to light scattering because they are 'hidden' from the light by other cells.

The standard curve shown in Figure 3.4 was obtained using a wavelength of 540nm. Wavelengths of between 420 and 650nm are commonly used to measure cell density. The sensitivity of the measurement increases sharply if light of a shorter wavelength is used. However, at lower wavelengths specific cell components, such as pigments, may absorb light and interfere with cell density measurement. Another practical consideration is absorbance by the culture medium. This should be measured and subtracted from absorbance values for the sample (culture medium plus cells). Alternatively, the spectrophotometer is calibrated so that it reads zero for medium not containing cells.

nephelometers   More sensitive instruments for measuring light scattering are called nephelometers. They have a light sensing device arranged at an angle to the incident beam of light and directly measure scattered light.

SAQ 3.3

The dry weight of a bacterial culture was measured by weighing. The culture was also diluted in culture medium and absorbances were recorded. The results are shown below:

Dry weight of predried filter = 2.7656g

Dry weight of filter plus cells from 10ml sample = 2.8456g

	Absorbance at 540nm
Medium minus cells	0.060
Undiluted culture	Infinity
Culture dilution:	
1 in 10	1.050
1 in 14	0.970
1 in 20	0.830
1 in 25	0.720
1 in 30	0.595
1 in 40	0.470
1 in 70	0.290

1) Plot a graph relating dry weight to absorbance.

2) Determine the dry weight concentration of a culture giving an absorbance (at 540nm) of 0.560 when diluted five-fold in culture medium.

3.4.3 Volume measurements

Volume can be used as a quick routine measurement to compare the amount of biomass in samples. Since biomass density varies little and assuming that the cell is 20% dry solids, a rough estimate of the cell mass can be made by dividing the cell volume by five.

packed cell volume

Volume measurements can be made by centrifugation of culture in a graduated centrifuge tube. The packed cell volume (PCV) obtained will depend not only on biomass concentration but also on the centrifugation speed and time. The centrifugation conditions are therefore usually fixed to minimise variability. The morphology of filamentous fungi also influence the PCV and prevents accurate determination of biomass from PCV measurements. Similarily, osmolarity of the medium will affect cell swelling and hence volume measurements.

3.5 Measuring cell components

There are circumstances where measurement of cell mass or numbers is not possible or impracticable. It is then often easier to measure the amount of certain cell components. This may be a macromolecular component, such as protein or DNA, or a low molecular weight molecule such as ATP. Such measurements are often applied for the estimation of biomass in the natural environment.

Π Can you think of circumstances where measuring cell constituents may be preferred over counting or weighing cells?

When cells grow as filaments or form clumps, counting or weighing are less suitable methods for measuring growth. Measuring cell constituents is also preferred for samples from natural habitats where non-cellular solids may interfere with cell counting or weighing methods.

The success of measuring cell components as a means of estimating biomass depends on:

- the absence of related materials in the non-cellular solids;

- the correlation of the cell component with cell mass.

protein
DNA and ATP

We will now consider three of the most commonly measured cell components for the estimation of biomass, namely protein, DNA and ATP.

3.5.1 Protein

protein

The protein content of cells does not vary greatly with different growth conditions, it therefore provides a reasonable estimate of cell mass. For example, 60-65% of the dry weight of bacteria is protein. The major limitation of protein analysis is the presence of non-cellular and water insoluble protein in medium components.

Biuret, Lowry
and Kjeldahl
methods

There are several methods of protein analysis that are useful for biomass estimation. Each will give a different protein value because they measure different properties of the protein. However, the values are meaningful if the basis of the analysis is understood. Two of the most widely used methods of protein measurement are based on the development of colour by reaction with chemical reagents; these are the Biuret and the Lowry methods. The Biuret method is easy, with good reproducibility since it measures peptide bonds, but it is not very sensitive. The Lowry method is more sensitive, but since it responds strongly to the aromatic amino acids errors are introduced unless samples and reference proteins are similar in composition. Another approach is the measurement of nitrogen content by the Kjeldahl method. This involves chemical digestion of the protein to liberate ammonia which is then measured by titration. The nitrogen content of biomass can be estimated to within 1% by this method. The value when multiplied by 6.25 gives a value for crude protein, but is subject to error from non-protein nitrogen. Typically the Biuret method can estimate protein in the mg range, the Lowry method is about 10-100 times more sensitive (that is, it can be used to measure samples containing about 10-100 µg protein ml^{-1}).

3.5.2 Deoxyribose nucleic acid (DNA)

DNA

The amount of DNA per cell varies with the growth rate. Faster growing cells have more DNA growing points and an increased amount of DNA. However, the variation of DNA per unit of cell mass is low and the DNA content of biomass, because of its constancy and specificity is correlated well with biomass. Despite this it has not been extensively used for estimation of biomass mainly because the methods are slower than those for protein.

The measurement of DNA in biomass is based on the estimation of deoxyribose by a colour reaction. The sensitivity of the method is similar to that for the measurement of protein by the Biuret method (ie it can be used to measure samples containing about

1mg DNA). Its measurement is useful for very complex culture media containing non-cellular insoluble protein.

Π Can you think of a reason why measurements of RNA, lipids and carbohydrates are not used for the estimation of biomass.

The main reason is that the relative concentrations of these cell components are influenced strongly by the growth conditions. Their correlation with cell mass is therefore poor.

3.5.3 Adenosine triphosphate (ATP)

All living cells contain ATP which is present in fairly constant amounts in each cell type and which is rapidly lost after the cells die. It follows that its concentration should be proportional to the concentration of viable cell mass. Measurement of ATP involves measurement of light produced as a result of an enzyme reaction catalysed by the enzyme firefly luciferase. The amount of light produced is proportional to the amount of ATP taking part in the reaction. The light producing reaction is:

firefly luciferase

$$ATP + luciferin + O_2 \xrightarrow{\text{luciferase}} oxyluciferin + AMP + PPi + CO_2 + light$$

In recent years, luciferin/luciferase reagents have been specially developed to give a constant light output with time. The light can be measured and recorded automatically using a luminometer. The procedure is very sensitive with light output being proportional to ATP concentration over the range 10^{-11} to 10^{-6} moles l^{-1}. In $E.\ coli$, for example, the ATP pool has been estimated to be 4.5-7.5 μ moles g^{-1} cell dry weight.

Π Use the information provided in the preceding paragraph to determine the minimum possible concentration of $E.\ coli$ required for biomass estimation measurement of ATP.

Minimum amount of biomass required = $10^{-11}/4.5 \times 10^{-6} = 2.2 \times 10^{-6}$ g l^{-1} cell dry weight.

Estimation of biomass by measurement of ATP assumes that the ATP content of the cell does not change during the sampling period and does not change with the growth environment. These assumptions, however, are difficult to justify and complicate the interpretation of data. For example, the ATP pool in $E.\ coli$ has been estimated to turn over 4-8 times per second. Therefore, an interruption of ATP synthesis during sampling could rapidly upset the ATP pool size and introduce large errors in biomass estimation.

Despite these potential sources of error, the luciferase assay has been used to estimate yeast and other microbial biomass in beer. Indeed, the correlation between the amount of cellular ATP and yeast cell growth has been shown to be good, using the luciferase assay.

Sections 3.3 to 3.5 have considered the measurement of cell numbers, cell mass and cell components. SAQs 3.4 to 3.6 are placed together to enable an indepth comparison of the advantages and limitations of the methods.

| **SAQ 3.4** | For each of the methods of biomass estimation listed 1) to 5), select factors which could greatly reduce the accuracy of the determination. |

Method of biomass estimation

1) Protein.

2) ATP.

3) Absorbance.

4) Wet weight

5) DNA.

Factors

Slow sampling speed.

Changes in biomass morphology.

A complex growth medium.

The presence of non-cellular particulate material.

Cell clumping.

Cell lysis.

| **SAQ 3.5** | The minimum dry weight of bacteria required for an estimation with an error of 2% are given below. Match these figures with one or more of the methods of biomass estimation listed. |

Minimum dry weight (mg)

50
10
1.0
0.1
0.001
0.00001

Method of biomass estimation

Biuret protein
Cell count
Lowry protein
DNA
Dry weight
Absorbance

SAQ 3.6	Identify each of the following statements as True or False. If False give a reason for your response.

1) An estimation of wet weight can be obtained by dividing the volume as determined by packed cell volume (PCV) measurement by five.

2) The Biuret method of protein determination measures peptide bonds in the protein molecules.

3) Measurement of ATP by bioluminescence is more sensitive than the measurement of protein by the Lowry method.

4) Wet weight measurements are usually more accurate than dry weight measurements.

5) The sensitivity of absorbance measurements increases as the wavelength of the light decreases.

6) A sampling time of around five seconds is desirable for ATP measurements on a growing culture of *E. coli*.

3.6 Measurement of nutrient consumption and product formation

carbon source, nitrogen source and oxygen

The amount of biomass formed can also be estimated indirectly from the amount of substrate consumed or product formed. The major nutrients which can be correlated with an increase in biomass are the carbon source, nitrogen source and oxygen. A generalised equation relating these nutrients to growth and product formation is:

C-source + N-source + $O_2 \rightarrow$ biomass + CO_2 + H_2O + products + heat

Each of the substrates and products in the reaction have been used for biomass estimation. The reliability of the approach depends on:

- the ratio of the biomass produced per unit of nutrient consumed or product formed (the growth yield coefficients - see also Chapter 4) ;

- the accuracy of measurement of the nutrient or the product.

Ideally there should be a direct proportionality, that is:

$X = Y_{xs}.S$ or $X = Y_{xp}.P$

where:

X = biomass concentration

Y_{xs} = growth yield coefficient for substrate (biomass produced per unit of substrate consumed)

Y_{xp} = growth yield coefficient for product (biomass produced per unit of product produced)

S = substrate concentration

P = product concentration

Also, the growth yield coefficients (Y_{xs} and Y_{xp}) should be constant. However, this may not always be true if the culture conditions vary and often the growth yield coefficients are dependent upon the growth rate.

The magnitude of the growth yield coefficient also influences the correlation of the measurement with increase in biomass. If it is low as for the carbon source (0.5-1.9 g cells g^{-1} carbon source), the nitrogen source (10 g cell g^{-1} nitrogen source) and oxygen (1.0 g cells g^{-1} O_2), then the correlation may be good. If the yield is high as for phosphate (30), sulphate (100) and magnesium (200), then a small error in substrate or product analysis will cause a large error in biomass estimation.

Large errors may also be introduced if the substrate is also used for product synthesis, particularly if the ratio of product to biomass is large and/or variable. This would have to be corrected for by independent measurement of product concentration.

SAQ 3.7   List six or more factors that could decrease the reliability of biomass estimation from measurements of nutrient consumption or product formation.

3.6.1 Carbon source consumption

carbon balance   Carbon is typically 47-53% of the dry cell weight. This suggests that measurement of the disappearance of the source of carbon should be a good means of following an increase in biomass. However, since the carbon source is often the energy source much of the carbon is converted to CO_2, and may also be converted to products. This balance can be described as follows:

$$C_{consumed} = C_{biomass} + C_{carbon\ dioxide} + C_{products}$$

It follows that to estimate biomass, carbon source consumption, CO_2 production and product formation must be measured independently. The accuracy of the measurement then depends on the accuracy of each method of analysis. Further complications are introduced when more than one major carbon source is present, for example, in a complex medium where glucose serves as the main energy source and amino acids, added as protein hydrolysates, are readily incorporated into biomass.

fermentation performance   Despite the limitations of the carbon balance method for estimating biomass, it is often an extremely useful approach for fermentation processes since the results may also be used to improve fermentation performance. For example, a knowledge of how carbon substrates are channelled in metabolism can lead to efficient control of carbon substrate concentration. This may improve the overall economics of the fermentation since the carbon source is usually the most expensive raw material in these processes.

3.6.2 Nitrogen source consumption

nitrogen source

Nitrogen is typically 8-12% of the dry cell weight. This suggests that an accurate measure of nitrogen consumption will give a reasonable estimate of the biomass. The most common source of nitrogen in culture media is ammonia, added as ammonia gas or ammonium salts eg $(NH_4)_2SO_4$. Other nitrogen sources are urea and amino nitrogen ($-NH_2$) from protein.

chemical analysis

ammonium selective electrode

When there are no nitrogen containing products formed, and ammonia is the major nitrogen source, its consumption is routinely and effectively used to estimate biomass in fermentations processed. Ammonia consumption can be measured by several methods, these include chemical analysis and the ammonium selective electrode. The main limitation of the chemical analysis is its difficulty to automate, whereas the ammonium electrode method suffers from interference by monovalent ions such as sodium and potassium.

3.6.3 Oxygen consumption

The importance of oxygen in aerobic cultivation of organisms has been discussed in previous chapters.

oxygen consumption

In cell cultivation, one of the major difficulties in the estimation of biomass is the requirement for continuous sterile sampling of liquid. However, in the case of oxygen consumption there is no such need since it can be determined from measurements of O_2 concentrations in outlet gases. This can be achieved by passing outlet gases to a mass spectrometer or a paramagnetic oxygen analyser.

mass spectrometry

Mass spectrometry is based on the separation of ionised molecules under vacuum. The separation, based on the mass to charge ratio, is achieved by magnetic instruments. Mass spectrometry has the advantage that it can be used to analyse for any one of several vapour phase components simultaneously. Compared to other methods of O_2 analysis it has a shorter response time (in the order of seconds rather than minutes) and has greater accuracy. Use of mass spectrometry has the added benefit of being able to be interfaced with a computer and can be used to monitor several fermentors. Presently mass spectrometers are finding increasing application in monitoring fermentations. This is despite its main disadvantage - cost - which is typically ten times more than a paramagnetic O_2 analyser.

paramagnetic oxygen analysis

Despite the advantages of mass spectrometers for analysing outlet gases, paramagnetic oxygen analysers are used extensively, primarily because of their lower cost. The analysis is based on attraction in a magnetic field since O_2 is strongly paramagnetic because it has two unpaired electrons. Compared with O_2, none of the other gases normally associated with bioprocesses are paramagnetic. Hence, this property can be used for selective oxygen analysis.

OTR: oxygen transfer rate

Estimation of biomass, by this approach, involves the determination of O_2 consumption rate by measurement of O_2 content in incoming and outgoing air. The exact oxygen transfer rate (OTR) is then determined by multiplying the aeration rate by the difference in O_2 content (inlet and outlet gases) and taking the absolute temperature and pressure into consideration. This is expressed mathematically as follows:

$$OTR = Q \, (C_{in} - C_{out})$$

where: Q = aeration rate;

C_{in} = oxygen content of inlet gas (air);

C_{out} = oxygen content of outlet gas;

OTR = oxygen transfer rate.

But the oxygen transfer rate is equal to the rate of consumption of oxygen by the culture of organisms (assuming that the concentration of oxygen in the medium remains constant). The rate of oxygen consumption will, of course, be proportional to the amount of biomass in the culture and to the rate at which these cells are carrying out metabolism. The rate at which the biomass will be using oxygen will be proportional to the rate at which cells are growing (ie their growth rate) and the amount of oxygen required to produce each unit of biomass. Thus to estimate the biomass concentration (X) of a culture, the growth yield coefficient for oxygen substrate (Y_o = g biomass g^{-1} O_2) and the specific growth rate (μ) must also be known.

$$OTR = \mu.X.(Y_o)$$

Then, $X = \dfrac{OTR}{\mu} \cdot Y_o$

where:

Y_o = growth yield coefficient for O_2 substrate (g of cells produced/g oxygen consumed);

Note that this equation holds providing OTR is expressed in units of g h^{-1}, μ in h^{-1} and Y_o in g g^{-1}. If OTR is expressed in mol.h^{-1} then we have to use the relationship.

$$X = \dfrac{OTR}{\mu} \cdot \dfrac{Y_o}{k}$$

where k = correction factor = 0.0313 mol O_2 g^{-1} O_2.

The specific growth rate, or the growth rate per unit amount of biomass, has units of reciprocal of time eg h^{-1}. We shall see in Chapter 5 that, in continuous culture where fresh medium is continually supplied to the fermenter, the specific growth rate equals the dilution rate of the culture. The important point to note, as far as this chapter is concerned, is that μ can often be determined readily and without prior knowledge of the biomass concentration.

Y_o can be determined using the following equation:

$$\frac{1}{Y_o} = 16 \left[\frac{2C + (H/2) - O}{Y_s \ M} + \frac{O'}{1600} - \frac{C'}{600} + \frac{N'}{933} - \frac{H'}{200} \right]$$

where:

C, H, O = Number of carbon, hydrogen, and oxygen atoms in substrate

C', H', O', N' = Percent of carbon, hydrogen, oxygen and nitrogen in biomass

M = Substrate molecular weight

Y_s = Growth yield coefficient for carbon substrate

We will not derive this equation here. Its derivation is however described in the BIOTOL text 'Bioprocess Technology: Modelling and Transport Phenomena'. Essentially what this equation states is that the Y_o value can be predicted from the chemical composition of the substrate and the composition of biomass.

Alternatively, Y_o may be determined directly by experiment, (see also Chapter 4).

SAQ 3.8	A continuous bacterial culture growing on glucose ($Y_s = 0.5$ g g^{-1}) was sparged with air at a rate of 1 l min^{-1}. Analysis of inlet and outlet gases showed that 0.5m moles O_2 l^{-1} air were removed by the culture. The elemental composition of the organism (percentage of dry weight) was mainly carbon (50), oxygen (20), nitrogen (14) and hydrogen (8). Determine the biomass concentration of the culture if the dilution rate(D) is 0.2 h^{-1}. (Glucose = $C_6H_{12}O_6$; relative molecular mass = 180). (assume that $\mu = D$ for this culture).

3.6.4 Carbon dioxide formation

carbon dioxide   Carbon dioxide is generally the most useful product to measure as a means of estimating an increase in biomass ie monitoring growth. It has value because:

- CO_2 is the most common catabolite;

- it does not require sterile sampling since CO_2 is in the gas phase;

- CO_2 can be easily measured accurately and rapidly.

As for O_2, CO_2 can be measured using a mass spectrometer. Indeed, determination of the CO_2 production rate also requires measurement of O_2 and gas flow rates.

The accuracy of biomass estimation by this approach relies on correlation between CO_2 production rate and growth. Since CO_2 is a terminal product of catabolism it reflects the generation of ATP. If the ratio of cell mass generated per mole of ATP is constant, and ATP is synthesised with constant efficiency, then CO_2 production rate is a good growth indicator. The exact ratio of CO_2 produced per unit of cell mass synthesised depends on factors such as the catabolic pathway used and the efficiency of coupling of energy generation (catabolism) with energy consumption (anabolism). The correlation will break down when growth and catabolism are uncoupled as is the case for cells in the stationary (non-growing) phase of the culture cycle.

Other catabolic products that are useful for monitoring fermentations are the end products of catabolism in anaerobic pathways. These include ethanol, lactate, acetate, propionate and butanol.

3.6.5 Heat formation

During microbial growth, heat is generated and often has to be removed to keep bioreactors (growth vessels) at constant temperature. The amount of heat generated is stoichiometrically related to rate of biomass formation (growth rate). It can therefore be used as a means of indirectly estimating microbial biomass.

In order to use heat of fermentation, it is necessary to consider the overall balance for the fermentation ie heat input and heat output. Heat input includes the heat generated during agitation (mixing) of the culture, as well as heat produced by the biomass. Similarly, heat output includes evaporation loss, conduction losses, loss in cooling system etc.

thermal yield The heat evolved per gram of biomass (thermal yield) will depend on the efficiency of the use of the energy substrate. Since the energy substrate is usually the carbon source, we can say that the thermal yield depends on the growth yield coefficient for carbon substrate. It follows that during fermentation in which the growth yield coefficient changes with time, the proportionality of heat to biomass will vary. For these fermentations, heat formation may not be suitable for the estimation of biomass.

SAQ 3.9

Using the lists provided, match each of the methods labelled 1) to 5), with appropriate merits and limitations.

Method of biomass estimation

1) Measurement of O_2 consumption.

2) Dry weight measurement.

3) Protein measurement.

4) Slide culture.

5) ATP measurement.

Merits

a) Avoids sterile liquid sampling.

b) Not influenced by non-cellular particulate material.

c) Highly sensitive and specific.

d) Distinguishes between viable and non-viable cells.

e) Result obtained within seconds.

Limitations

i) Does not distinguish between viable and non-viable cells.

ii) Requires special and expensive equipment.

iii) Greatly influenced by soluble components of culture medium.

iv) Lacks sensitivity.

v) Result obtained only after several hours.

Summary and objectives

In this chapter we have seen that there are four fundamentally different approaches to the estimation of biomass. These are: measurement of cell numbers; measurement of cell mass; measurement of cell components; measurement of nutrient consumption and product formation. For each of these approaches there are several different methods and choice of an appropriate method depends on a variety of factors. These include: the nature of the biomass, the sensitivity, accuracy and speed of the measurement; the presence and nature of non-cellular contaminants; the availability of equipment; the possible need to distinguish viable and non-viable cells; the possible need to avoid liquid sampling of culture.

Now that you have completed this chapter you should be able to:

- list merits and limitations of the various approaches to the estimation of biomass;

- select an appropriate method of estimating biomass according to the nature of the sample;

- describe a variety of different methods of estimating biomass;

- determine biomass concentration from data derived from a variety of different methods;

- list the main sources of errors associated with different methods of biomass estimation.

Growth in batch culture

Growth in batch culture

4.1 Introduction

Microbial growth is the orderly increase in all cellular constituents, which ultimately results in the formation of new cells. Most bacteria divide into two daughter cells by transverse binary fission. Before dividing, a cell must reproduce all of its essential constituents and double its mass. This sequence of growth and division continues as long as the cells are in an environment that provides the essential nutrients and physical conditions needed for growth.

In this chapter we shall be dealing with growth of micro-organisms in closed systems, so-called batch cultures, where all the nutrients are present in the medium before inoculation and no additions are made to the medium after inoculation.

phases of the growth cycle in batch culture

Growth in a batch culture shows distinct phases. At first growth is usually not apparent or slow (lag phase), after which exponential growth sets in with a steep increase in cell numbers/biomass (exponential phase). Sooner or later during this phase toxic metabolites will build up and the nutrient resources will become exhausted with the result that the rate of growth declines and finally comes to a halt (stationary phase). When the energy reserves of the cells themselves are depleted they will eventually die (death phase). The time these phases take very much depends on the organism involved and the circumstances under which it grows.

4.2 Growth

definition of growth

In general growth is defined as the orderly increase in all chemical cell components and structures. In this sense increase in biomass does not necessarily imply growth, since part of the mass may be there only temporarily as storage products, such as glycogen or poly-β-hydroxybutyrate. Similarly an increase in cell number is not necessarily orderly since at the moment of cell division a discontinuity occurs when one cell splits in two, although in that case the total microbial mass as well as the activity of certain cell components (for instance enzymes) do increase in an orderly way.

Having defined growth, the growth rate for unicellular systems is defined as the increase of cell number or mass per unit time. The new cell grows to its mature size (about double its original size) and then divides into two cells (binary fission), which in turn grow and divide. Each time a cell divides it is called a generation and the time it takes, the generation time. In fact the generation time is the time lapse in which the population doubles, which is why it is also known as doubling time. After one generation time both cell mass and cell number have doubled. Generation times vary between micro-organisms: the majority of bacteria take one to three hours to double in cell number but there are also some that only need 10-20 minutes and other microbes that take 24 hours or more (some protozoa and algae).

Since it is very difficult to determine the growth of a single cell, in microbiology growth always refers to the increase in cell material of a population. A population is defined as a group of individual cells of the same strain, this means a group of individual cells derived from one mother cell.

balanced growth

When the micro-organisms are fully adapted to a medium that is adequate for their requirements, the cells will be in a state of balanced growth. In this state an increase in biomass implies a comparable increase of all measurable cell components of the population, that is protein, DNA, RNA and intracellular water content. This means that cultures in a state of balanced growth maintain a constant chemical composition. This adds a third way of determining the growth rate of a population: measuring the increase of any one component is representative of the population's growth.

asynchronous and synchronous growth

It is important to realise that even during balanced growth not all of the cells will be dividing at precisely the same time; this is termed asynchronous growth. When all cells in a population (culture) divide simultaneously the culture undergoes synchronous growth. We shall see later in this chapter that synchronous growth of laboratory cultures can be achieved using a variety of methods, although it lasts only a few generations.

Bacteria can be grown in a closed system, known as a batch culture or in an open system, where growth is continuous. In a continuous (open) culture growth is controlled by the addition of fresh nutrients and removal of spent medium and cells from the vessel; this will be considered in the next chapter (Chapter 5).

4.2.1 Exponential growth in batch cultures

At the start of a batch culture all necessary nutrients are present in certain concentrations. During the time of culturing nothing is added to or taken from the culture so that the nutrient concentrations change only because of the activity (growth) of the cells. For instance growth of a bacterium in a conical flask is a batch culture and so is a culture growing on solid medium in a covered petri dish.

semi-logarithmic plots

Binary fission can, in favourable conditions, lead to exponential growth: the number of cells increases by a constant factor per unit of time. One of the characteristics of this type of growth is that the rate of increase in cell number escalates at an ever faster rate.

When the exponential growth of a population is graphed on arithmetic coordinates (time and cell number) the result is a curve with a constantly increasing slope (Figure 4.1). This provides us with a rather inconvenient representation of growth which is why usually the logarithm of the data is taken. Figure 4.1 also shows the result of plotting the logarithm of the cell number against the arithmetic value of elapsed time (a so-called semi-logarithmic graph). It is a straight line, which is much easier to interpret.

In Figure 4.1 logarithms to base 10 ($\log_{10}$) are used to produce the straight line. This base is often used in growth studies because it generates numbers that are relatively easy to work with and interpret. However, $\log_2$ is sometimes used because it describes growth by binary fission ie cell division produces a two fold increase in cell number. Similarly, logarithms to base e or natural logarithm ($\log_e$ or 1n) are commonly used since it closely resembles actual growth in batch culture ie it takes into account the fact that cells are dividing at different times within the cell division cycle (asynchronous cultures). Regardless of the base used, exponential growth is converted to a straight line by taking logarithms of the growth data.

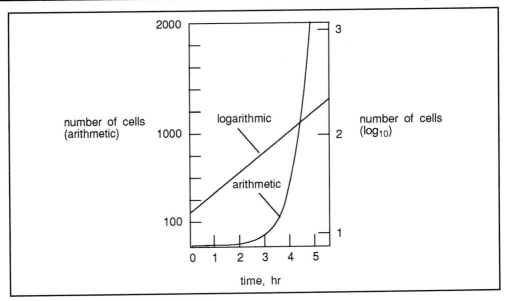

Figure 4.1 Growth of a microbial batch culture.

Mathematically exponential growth can be described in two ways. One approach takes the number of cells as a starting point (N_o). The change in number caused by binary fission, can then be expressed in a simple equation.

For example, starting from one cell ($N_o = 1$)

$$1 \xrightarrow{(1)} 2 \xrightarrow{(2)} 4 \xrightarrow{(3)} 8 \xrightarrow{(4)} \cdots \xrightarrow{(n)} $$

$$N_o \qquad N_o.2 \qquad N_o.2.2 \qquad N_o.2.2.2 \qquad N_o.2^n$$

So the number of cells after a period of time (N_t) in an exponentially growing population can be expressed as:

$$N_t = N_o.2^n \qquad\qquad\qquad (E - 4.1)$$

Where n is the number of divisions and, written in its logarithmic form:

$$\log N_t = \log N_o + n \log 2 \qquad\qquad\qquad (E - 4.2)$$

The number of divisions (n) is therefore:

$$n = \frac{\log N_t - \log N_o}{\log 2} \qquad\qquad\qquad (E - 4.3)$$

and the number of divisions per time unit

$$\frac{n}{t} = \frac{\log N_t - \log N_o}{t \log 2} \qquad\qquad\qquad (E - 4.4)$$

The quotient n/t is called the division rate constant (k); this can be considered as an average value for the population over a finite period of time. Another way of considering the division rate k is the average number of generations occurring per hour.

doubling or generation time

Often we are not interested in the number of divisions per time unit but rather in the doubling time (t_d) or generation time, (t_g). This parameter can be expressed as:

$$t_d \ = \ \frac{t}{n} \ = \ \frac{1}{k} \qquad\qquad\qquad (E - 4.5)$$

∏ In a period of 5 hours the number of cells in a batch culture increases from 10^3 to 10^6. Determine the value of both k and the doubling time (t_d).

Using Equations 4.4 and 4.5 the division rate constant is:

$$k \ = \ \frac{\log 10^6 - \log 10^3}{5\log 2} \ = \ \frac{3}{1.505} \ = \ 2h^{-1}$$

which means there are two divisions per hour and the doubling time is 0.5 hour.

Another approach is to describe the growth of a culture not by taking the individual cell as starting point but to consider the growth of the population as an autocatalytic reaction. The rate of a catalysed reaction depends on the amount of catalyst present. In this case the biomass itself is the actual catalyst, and the rate of biomass production depends on the amount of biomass (X) present at any particular time.

$$\frac{dx}{dt} \ = \mu \, . \, X \qquad\qquad\qquad (E - 4.6)$$

growth rate and specific growth rate

In this equation, dx/dt represents the increase in biomass in a certain period of time ie the growth rate, and μ the (momentary) growth rate constant or specific growth rate. Conversion of Equation 4.6 results in the following equation:

$$\mu \ = \ \frac{1}{X} \, . \, \frac{dx}{dt} \qquad\qquad\qquad (E - 4.7)$$

In other words the specific growth rate $= \dfrac{\text{rate of growth}}{\text{amount of biomass}}$

Integration of Equation 4.7 gives the biomass of an exponentially growing culture at any particular time:

$$X_t = X_o \, . \, e^{\mu t} \qquad\qquad\qquad (E - 4.8)$$

and, after taking the natural logarithm,

$$\mu \ = \ \frac{\ln X_t - \ln X_o}{t} \qquad\qquad\qquad (E - 4.9)$$

⊓ Use the equations to establish the relationship between the doubling time (t_d) and the specific growth rate (μ) (Attempt this before reading on).

This can be established in more than one way. Method 1:

$$\mu = \frac{\ln X_t - \ln X_o}{t} \quad \text{(Equation 4.9)} \qquad t_d = \frac{t}{n} \quad \text{(Equation 4.5)}$$

Combining Equation 4.5 with an expression for n Equation 4.3, we have:

$$t_d = \frac{t \log 2}{\log N_t - \log N_o}$$

If we right this in terms of biomass, rather than cell numbers:

$$t_d = \frac{t \ln 2}{\ln X_t - \ln X_o}$$

Solving for t:

$$t = \frac{\ln X_t - \ln X_o}{\mu} \text{ from Equation 4.9 but also } t = \frac{t_d \ln X_t - \ln X_o}{\ln 2}$$

Combining expressions for t:

$$\frac{\ln X_t - \ln X_o}{\mu} = \frac{t_d . \ln X_t - \ln X_o}{\ln 2}$$

$$t_d = \frac{\ln X_t - \ln X_o}{\ln X_t - \ln X_o} . \frac{\ln 2}{\mu}$$

$$t_d = \frac{\ln 2}{\mu} \text{ ie } \frac{0.693}{\mu}$$

Method 2:

To find the doubling time we assume $X_t = 2X_o$ at time $t = t_d$ and transfer this to Equation 4.8:

$$2X_o = X_o . e^{\mu t_d}$$

After taking the natural logarithm $\mu t_d = \ln 2$ or $t_d = 0.693/\mu$.

If we compare $t_d = 0.693/\mu$ with Equation 4.5 we find the relationship between the division rate k and the specific growth rate (μ):

$$\mu = 0.693 \, k.$$

There is an essential difference between μ and k: μ only refers to the growth rate at a certain moment, and can be different before or after, whereas k is an average value for the population over a period of time.

| SAQ 4.1 | 1) A bacterial culture containing 100 cells increased in population to one billion cells (10^9) in 10 hours. Determine (in suitable units): |

1) A bacterial culture containing 100 cells increased in population to one billion cells (10^9) in 10 hours. Determine (in suitable units):

   a)   the number of generations (n)

   b)   the generation time (t_g)

   c)   the division rate constant (k)

   d)   the specific growth rate (μ)

   e)   the growth rate at the end of the incubation.

What assumption do you have to make for your determinations to be valid?

2) A bacterial culture containing 100 cells has a generation time of 15 minutes. How long will it take for this culture to reach a population of one million cells?

4.3 The growth cycle

the growth cycle

Growth of a microbial population in batch culture is only partly represented by the exponential curve in Figure 4.1. Growth starts after inoculation of an appropriate growth medium with cells from a starter culture and incubation in favourable conditions. A typical representation of the growth cycle that follows is presented in Figure 4.2. The growth curve can be divided into several distinct successive phases: lag phase, exponential phase, stationary phase, and death phase. It must be emphasised that these phases are reflections of the events taking place in a population and that they do not apply to individual cells.

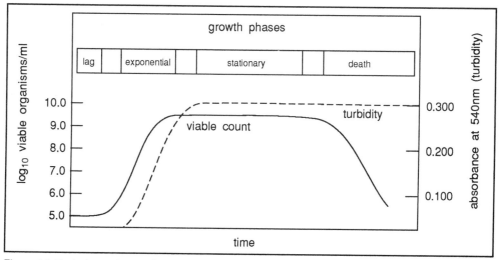

Figure 4.2 Typical growth curve for a bacterial population. Adapted from Brock, T.D, W. and Madigan, MT (1984) Biology of Micro-organisms 5th Edition, Prentice Hall.

4.3.1 Lag phase

lag phase

In the lag phase, the microbial population does not show any perceptible growth: there is no increase in cell number or turbidity as the cells are adapting themselves to their new environment. This can be due to a number of factors. The medium may not be optimal for the organism so enzymes may need to be synthesised to be able to use the substrate as energy source or for the synthesis of cell material. The medium may contain toxic compounds which need to be removed. If spores are used for inoculation these need to germinate first before there can be any actual increase in cell number or mass. In all these instances the lag phase may be relatively long.

∏ Can you think of a circumstance in which a lag phase does not occur?

If cells from a culture in the exponential phase are transferred to the same medium, growth can continue unhampered, since the cells are completely adapted to that medium.

The lag time also depends on the inoculum itself. As we have seen, if it is taken from a population in the exponential phase there may not be a lag phase at all, but when taken from an older culture (in the stationary phase) many of the cells needs to resynthesise cell components for division and need to replenish their pool of coenzymes and other essential metabolites. Therefore, in the lag phase the mRNA content of the cell is relatively high.

As cells become adapted to the growth medium they begin to divide. The division rate (k; see Section 4.2.1) is initially slow but increases until the culture enters the exponential phase of growth, where k remains constant. The period of increasing division rate, between the lag and exponential phases of growth, is often referred to as the acceleration phase of a batch culture.

4.3.2 Exponential phase

exponential phase

In Section 4.2.1 we saw that exponential growth is typical for growth by binary fission and that the rate of exponential growth can vary between micro-organisms. To mention two extremes: *Salmonella typhi*, which causes typhoid has a generation time of 20 to 30 minutes when grown in culture, whereas the tubercle Bacillus *Mycobacterium tuberculosis* needs a whole day to double once or twice. In general one can say that prokaryotes grow faster than eukaryotic micro-organisms and that of the latter group the smaller ones grow faster than the large ones. The rate of exponential growth depends also on external factors, such as temperature and composition of the culture medium.

Microbial populations seldom maintain exponential growth at high rates for a long time. The reason for this is obvious if one considers the consequence of exponential growth.

∏ Presume exponential growth of a single bacterium (weight about 10^{-12}g) for 48 hours, with a doubling time of 20 min. What is the weight of the progeny produced?

The number of new cells can be calculated using Equation 4.1:

$$N_t = N_o 2^n$$

We know that $N_o = 1$ and from Equation 4.5 $t_d = \dfrac{t}{n}$

This can be rearranged to:

$n = \dfrac{t}{t_d}$ thus:

$N_o = 1$ and $n = \dfrac{t}{t_d}$

$n = \dfrac{48}{0.3333} = 144$

$N_t = 1.2^{144} = 2.23 \times 10^{43}$

The mass of the progeny would be 2.23×10^{28} kg or roughly 4000 times the weight of the Earth!

4.3.3 Stationary phase

stationary phase

In batch cultures the exponential phase cannot go on for ever. There comes a moment when either the nutrient source is depleted or toxic metabolic products have accumulated to such a level as to inhibit growth. This gives at first a deceleration period of growth which leads to a phase where there is no further increase in viable cell number, the so called stationary phase. This does not necessarily mean that some cells are still dividing whilst others are dying. This is however often the case for cyanobacteria and algae. In the majority of cases there is a certain rest stage, no longer reproductive. If cells in the stationary phase are inoculated in a fresh medium growth starts again after a certain lag phase. During the transition (deceleration) period between the exponential phase and the stationary phase, population growth is unbalanced; this means the various cell components are synthesised at an unequal rate. As a result the composition of cells in the stationary phase is different from those in the exponential phase. The actual composition depends on the growth-limiting factor. Some of the components that are left in the medium can still be useful for the bacteria although they are not essential. Metabolism of these compounds leads to the so-called secondary metabolites (as opposed to the primary metabolites, formed during the exponential phase). These secondary metabolites are important for industry, with antibiotics as the best-known example.

In general one can say that cells in the stationary phase are smaller than those in the exponential phase and they are more resistant to adverse physical conditions such as heat, cold, radiation, and chemical agents.

4.3.4 Death phase

death phase

Eventually, if bacteria are kept in the stationary phase, in the same medium, the population will enter the death (decline) phase. Death results from a number of causes, of which depletion of cellular energy reserves is an important one.

During this phase the total cell count may remain the same but the viable count decreases. Like growth, death (measured by the difference between the total cell number and the number of viable cells) also occurs exponentially, as can be seen in Figure 4.2. The death rate of bacteria varies according to circumstances and also from

---

organism to organism. In some cases death is accompanied by cell lysis, leading to a decrease in total count.

SAQ 4.2

After inoculation of a bacterial culture you measure its absorption in a spectrophotometer and find the value is 0.005. The next measurement is done after 16 hours, and after that you measure every hour. You find the following absorbance (at 540 nm) values: 0.040; 0.064; 0.101; 0.160; 0.249; 0.286; 0.300; 0.302.

1) Draw the growth curve of this culture by making a semi (natural) logarithmic plot of the data.

2) What is the specific growth rate (μ) and the doubling time (t_d) of the culture in the exponential phase?

3) Estimate the length of the lag phase.

4.4 Effect of nutrient concentration on growth

growth limiting nutrient

Nutrient concentration can affect growth in two ways. Firstly it can limit the total amount of biomass, the maximum crop, and secondly it can influence the growth rate. The nutrient, within a growth medium, that influences growth in this way is called a limiting nutrient.

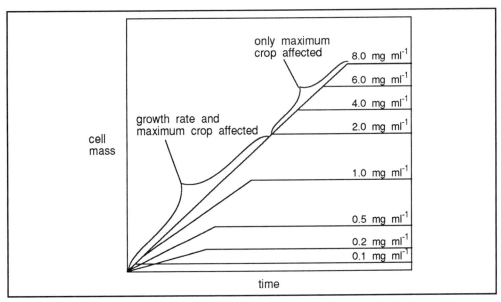

Figure 4.3 Relationship between nutrient concentration, maximum crop and growth rate (see text).
Adapted from Brock, T.D, and Madigan, MT (1988), Biology of Micro-organisms 5th Edition, Prentice Hall.

Figure 4.3 shows that at low nutrient concentrations both maximum crop and the growth rate are affected. That less nutrient results in less total growth is easy to understand since part of the nutrients are used to build cell components. The limiting effect on the growth rate can be explained in a different way. The effect is only noticed when the concentration of the limiting nutrient is sufficiently low. Figure 4.4 shows the curve representing the relationship between the limiting nutrient glucose and the growth rate of *Escherichia coli*. Note that a concentration of 10 mg ml^{-1} would give the same response as 8 mg ml^{-1} indicates that growth rate is not affected at higher substrate concentrations.

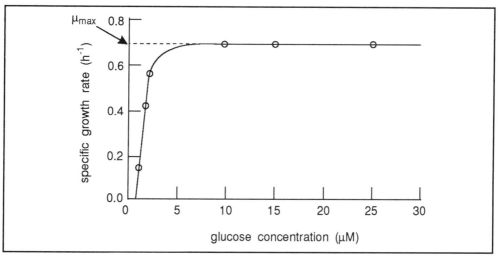

Figure 4.4 The relationship between the concentration of the limiting nutrient, glucose and the specific growth rate of *Escherichia coli*. Data taken from Stanier, R.Y, Ingraham, J.L, Wheelis, M.L and Painter, P.R. (1986) General Microbiology. MacMillan Basingstoke, UK.

The curves relating specific growth rate and nutrient concentration are typically hyperbolic and resemble a saturation process of the kind also seen for active transport and enzyme action. Probably, at very low external concentration of the limiting nutrient, the nutrient cannot be transported into the cell fast enough to fulfil the metabolic demand for it. It may be that not all carrier sites of the permease system (transport proteins) are occupied. At higher concentrations permeases are capable of maintaining saturating intracellular concentrations of the limiting nutrient, resulting in a maximal growth rate. In many cases the curve fits the so called Monod equation:

$$\mu = \mu_{max} \frac{[S]}{K_s + [S]}$$

Monod constant

in which μ is the specific growth rate at limiting nutrient concentration ([S]) and μ_{max} is the specific growth rate at saturating concentration. K_s is a constant, analogous to the Michaelis-Menten constant (K_M) of enzyme kinetics, and is called the Monod-constant. Numerically, K_s is equal to the substrate concentration that supports a growth rate equalling $\frac{1}{2}$ μ_{max}. K_s values for *Escherichia coli* growing on glucose are of the order of 1 x 10^{-6} M or 0.18 milligram per litre (see Figure 4.4). These very low values reflect the high affinities which are characteristic of many bacterial permeases. These high affinities can be interpreted as an evolutionary adaptation to growth in extremely dilute solutions. In this respect conventional laboratory media are very different from natural circumstances.

The ability to alter the growth rate by varying the availability of nutrient is used for the operation of a device for continuous growth, a so-called chemostat, which is discussed in the next chapter (Chapter 5).

Π In which phase of growth in a batch culture will the growth rate be limited by a low concentration of one of the nutrients?

The growth curve (Figure 4.2) shows a decrease in the growth rate of the culture at the end of the exponential phase (the deceleration phase). This may be caused by a decrease of the concentration of the limiting nutrient below the saturation level which permits growth at $\mu_{max.}$

SAQ 4.3

Identify each of the following statements as True of False. If False give a reason for your response.

1) The growth rate, $\frac{dx}{dt}$ is constant in exponential phase of growth.

2) The growth rate increases with decreasing generation time.

3) $\frac{dx}{dt}$ always equals zero in the stationary phase of the bacterial growth cycle.

4) K_s determines residual concentrations of limiting nutrient after growth in batch culture.

5) The maximum specific growth rate can be determined from the slope of a semi-logarithmic plot for the exponential phase.

6) K_s determines the specific growth rate at saturating high nutrient concentrations.

The (net) amount of growth of a bacterial culture is the difference between the biomass of cells (or the cell number) as inoculum and the cell mass present in the culture when it enters the stationary phase. This is the growth yield (Y).

When growth is limited by one particular nutrient there is a fixed relationship between the net growth and the concentration of that nutrient (see Figure 4.5). The mass of cells produced per unit of limiting nutrient is called the growth yield coefficient Y_s and can be calculated using the equation:

growth yield coefficient

$$Y_s = \frac{X - X_o}{S}$$

in which:
X = dry weight of cells (eg in mg l^{-1}) at the beginning of the stationary phase
X_o = dry weight of inoculum (in mg l^{-1})
S = concentration of limiting nutrient (in mg l^{-1}).

The slope of the line in Figure 4.5 represents the growth yield coefficient (Y_s). In this way the growth yield can be determined for any of the required nutrients. Conversely, the concentration of a limiting nutrient in an unknown, otherwise complete, medium can be calculated by measuring the total growth of a bacterium on it. This method is known as a bioassay and has been widely used for the determination of concentrations of amino acids and vitamins in food. Although in some applications it has been replaced by other

bioassay

(chemical and physical) methods it is still important and much used for detection and quantitation of compounds involved in growth. The only pre-requisite is a microbial strain for which the compound to be assayed is an essential nutrient.

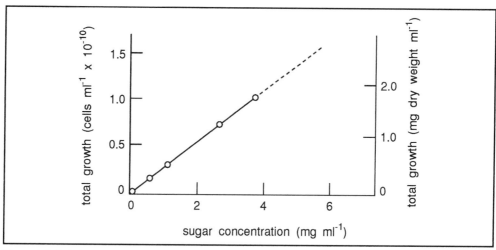

Figure 4.5 Relationship between net growth of an (aerobic) bacterium (*Pseudomonas* sp.) and the initial concentration of the limiting nutrient (fructose).

∏ Calculate the growth yield coefficient on fructose for a *Pseudomonas* sp. from the data presented in Figure 4.5.

The growth yield coefficient can be read from the slope of the line in Figure 4.5, and is approximately 0.5 g biomass (g^{-1} substrate).

molar growth yield
The growth yield can also be expressed as the number of grams (dry weight) of biomass formed per mole of substrate utilised. This is the so-called molar growth yield (Y_{molar}). In the case of the *Pseudomonas* sp. growing on fructose (Figure 4.5), the growth yield coefficient is 0.5 g g^{-1} and the molecular mass of fructose is 180, so $Y_{molar} = 0.5 \times 180 = 90$ g $mole^{-1}$. We shall see in the next section (energetics of growth) how Y_{molar} values can be used.

metabolic quotient
The metabolic quotient (q) is another parameter which can be used to describe the performance of cultures in relation to substrate utilisation. The rate of utilisation of substrate (s) at a particular moment is given by:

$$\frac{ds}{dt} = q\,X$$

so, $q = \dfrac{ds}{dt} \cdot \dfrac{1}{X}$

This expression for q indicates that the metabolic quotient is the rate of substrate utilisation per unit amount of biomass (or the specific rate of substrate utilisation).

The metabolic quotient can also be expressed in terms of specific growth rate (μ) and growth yield coefficient (Y_s).

Since $\mu = \dfrac{dx}{dt} \dfrac{1}{X}$ (see Equation 4.7) and $Y_s = \dfrac{\text{amount of biomass produced}}{\text{amount of substrate used}} = \dfrac{dx}{ds}$

then:

$$\frac{\mu}{Y_s} = \frac{dx}{dt} \cdot \frac{1}{X} \cdot \frac{ds}{dx} = \frac{ds}{dt} \cdot \frac{1}{X}$$

so, $q = \dfrac{\mu}{Y_s}$

This equation can be used to estimate demands for substrate at different growth rates.

∏ The specific growth rate (μ) of a culture in exponential phase is 0.2 h^{-1} and the growth yield coefficient (Y_s) for limiting nutrient in 0.4 g biomass (g^{-1} substrate). Determine the metabolic quotient for limiting nutrient in suitable units? Would you expect q to change during exponential growth? Give a reason for your response.

$q = \dfrac{\mu}{Y_s} = \dfrac{0.2}{0.4} = 0.5$ g substrate h^{-1} (g^{-1} biomass)

No, q would remain constant during exponential growth because both μ and Y_s are constant.

SAQ 4.4

For a bacterial strain grown on glucose:

1) Determine the growth yield coefficient (Y_s) from the following data.

Glucose concentration (g l$^{-1)}$)	Dry weight biomass produced (g l^{-1})
0.5	0.25
1.5	0.75
2.5	1.25
3.5	1.75

Assume all of the substrate is used.

2) If the metabolic quotient for glucose ($q_{glucose}$) is 2.5g h^{-1} (g^{-1} biomass), what is the specific growth rate (μ) in suitable units?

4.5 Energetics of growth

efficiency of growth

When considering growth of chemoheterotrophic micro-organisms the growth yield, when measured in relation to the quantity of organic substrate utilised, is an index of the efficiency of converting substrate into biomass. For example, the graph shown in Figure 4.5 was obtained for an obligately aerobic chemoheterotrophic pseudomonad growing on a synthetic medium containing only one nutrient (fructose) as carbon and energy source. It shows a growth yield coefficient of about 0.5 g g^{-1}. The carbon content of fructose is about 40% and that of cell constituents 50%. So the fraction of fructose carbon that is actually converted to biomass carbon is approximately 0.62 ie:

$$\left(\frac{50}{40} \cdot 0.5\right) \times 100 = 62\%.$$

Presumably the cell uses just over half of the fructose to make cells and oxidises the other portion to carbon dioxide. Similar experiments with other aerobic chemoheterotrophs show an efficiency for the conversion of nutrient carbon to cell carbon usually falls between 20 and 50 per cent. These differences may be caused by differences in efficiency in generating ATP by dissimilation of the substrate.

energy
requirements
for growth

Growth is an energy-requiring process. It involves investment of energy in the formation of building blocks (monomers), in polymerisation of monomers and in processes such as active transport and movement. The energy requirements of monomer and polymer synthesis can be calculated from their known biosynthetic pathways. However, the amount of energy used for active transport and other (mechanical) processes cannot be estimated, but it is probably small compared to that required for biosynthesis.

If micro-organisms are grown on rich culture media, in which all monomers are likely to be provided the energy required serves mainly for polymerisation. Knowing the macromolecular composition of the cell and the amount of energy needed for polymerisation, the energy spent on polymer synthesis (in moles ATP per monomer and per quantity of cells) can be calculated. This has been calculated for several macromolecules. (Table 4.1).

Substance	Monomer units in polymer			ATP required	
	Approximate dry weight, per cent	Average molecular mass	Micromoles/ 100 mg cells	Per monomer	Micromoles ATP/100 mg cells
Protein	60	110	545	5	2,725
Nucleic acid	20	300	67	5	335
Lipid	5	262	19	1	19
Polysaccharide	5	166	30	2	60
Peptidoglycan	10	1000	10	10	100
Total					3,239
					or 31 g cell material per mole ATP

Table 4.1 Energy expenditure for polymer synthesis. Data derived from Brock, T.D and Madigan, MT (1988) Biology of Micro-organisms, Prentice Hall.

Two things need to be taken into account when reading this table. Firstly, the values presented are only approximate since the cell composition varies among organisms and according to growth conditions. Secondly, not all energy requiring polymerisation steps may be known yet. The figures indicate that for every 100 mg dry mass of cells formed, more than 3000 µmoles of ATP are needed for polymerisation. In other words, from one mole of ATP a theoretical maximum of 31 gram cell material (dry weight) can be formed if no energy is spent in the formation of monomer building blocks. It is also obvious from Table 4.1 that protein needs the most polymerisation energy.

Π   The protein content of a bacterial culture, growing on an amino acid rich medium, was shown to be 56% of cell dry weight. Assuming that the average molecular weight of amino acids is 110 and that 5 ATP's are required to polymerise each monomer, determine: 1) Micromoles monomer per 100 mg cell dry weight. 2) Micromoles of ATP consumed in making protein per 100 mg cell dry weight produced.

1)   Micromoles monomer per 100 mg cell dry weight. Since 56% of the cell dry weight is protein 100 mgs of cells contains 56 mg of protein. But 56 mg protein contains $\frac{56}{110}$ millimoles of monomer if the average molecular weight of amino acid is 110.

$$= \frac{56}{110} \cdot 1000 \; \mu\text{moles} = 509 \; \mu\text{moles}$$

2)   Micromoles ATP consumed to make the protein in 100 mg cell dry weight = Micromoles monomer per 100 mg cell dry weight x ATP required for monomer polymerised = 509 x 5 = 2545 µmoles ATP.

Y_{ATP}

The actual energy costs of growth may be examined by measuring the cell yield (biomass produced) in comparison to the amount of ATP generated during catabolism, a parameter which is known as Y_{ATP}. Determination of Y_{ATP} is most readily carried out on fermentative micro-organisms, there are two reasons for this:

• the amount of ATP produced per molecule metabolised substrate is well known;

• during growth on rich medium the carbon source is used solely for the generation of energy, other constituents of the medium (such as amino acids) are used for biomass formation.

This is not the case for respiring organisms for which it is not known exactly how much of the energy generated is used for ATP generation and how much is used for other purposes.

Y_{ATP} can be determined by culturing the organism anaerobically in a rich medium which contains all required monomers and a single fermentable electron donor such as glucose, in known concentration. At the end of growth the amount of glucose consumed is determined and the dry weight formed during growth is measured. It is necessary to know the pathway of glucose breakdown to calculate the amount of ATP produced. It should be noted that this approach mainly measures the energy required for polymer formation.

Table 4.2 summarises the results of Y_{ATP} determinations for various micro-organisms. From this table it is obvious that cell yields per mole of substrate (Y_{molar}) vary markedly amongst micro-organisms. The observed differences are a reflection of the different fermentation pathways, each with their own ATP yield. For instance *Zymomonas mobilis* ferments glucose to ethyl alcohol via the Entner-Doudoroff pathway with an ATP yield of 1 mole per mole of glucose. In contrast, yeasts produce twice as much biomass than *Z. mobilis* per mole of glucose fermented but also generate double the amount of ATP.

Organism	Y_{molar} (grams dry weight per mole substrate used)	ATP yield (moles ATP per mole of substrate)	Y_{ATP} (grams dry weight per mole ATP)
Streptococcus faecalis	20	2	10
Streptococcus lactis	19.5	2	9.8
Lactobacillus plantarum	18.8	2	9.4
Saccharomyces cerevisiae	18.8	2	9.4
Zymomonas mobilis	9	1	9
Aerobacter aerogenes	29	3	9.6
Escherichia coli	26	3	8.6

Table 4.2 Growth yield of fermentative micro-organisms, measured in terms of glucose fermented (Y_{molar}) and ATP produced (Y_{ATP}). Data derived from Brock, T.D and Madigan, MT (1988) Biology of Micro-organisms, Prentice Hall.

It is also clear from Table 4.2 that Y_{ATP} values are quite similar for different organisms and have an almost constant value between 9 and 10. The fact that Y_{ATP} is more or less constant can be used to determine the ATP yield of an unknown dissimilation route.

∏ We have seen that from 1 mol ATP a theoretical maximum of 31 g biomass (dry weight) can be formed. However, it appears that fermentative micro-organisms synthesise only around one-third of this amount. Can you explain this?

The difference probably reflects our lack of knowledge of some energy-requiring processes (active transport, motility, etc) but can also be attributed to wastage of energy by the cell. Furthermore, a certain amount of energy is needed to maintain cell structure and integrity (called maintenance energy, see also Chapter 5). Substrate used for this purpose is not available for cell growth.

SAQ 4.5

A bacterial culture can use methanol (CH_3OH), ethanol (CH_3CH_2OH) or methane (CH_4) as sole source of carbon and energy. Biomass yield values for the growth substrates were 0.4, 0.3 and 0.8 g g^{-1} respectively. Express these values as Y_{molar}. What are the carbon conversion efficiencies if the cell dry weight is found to contain 45% carbon (atomic mass: carbon = 12; oxygen = 16; hydrogen = 1)?

1) Imagine you are cultivating a yeast in anaerobic conditions in batch culture on a rich medium, in which the glucose concentration is growth limiting (20 g l^{-1}). After 50 hours exponential growth came to a halt and you harvest the yeast. The biomass concentration and residual substrate concentrations, appear to be 2 g l^{-1} and 0.25 g l^{-1} respectively. During cultivation you monitored growth with a spectrophotometer and found that the specific growth rate (μ) was 0.1 hr^{-1}. For inoculation you used a yeast suspension that was precultivated in the same medium and had reached the exponential phase. (Molecular mass of glucose, $C_6H_{12}O_6$ is 180 and 2ATP per mole of glucose is generated via glycolysis).

   a) After how many generations have you harvested and how much yeast did you use for inoculation?

   b) What is the molar growth yield (Y_{molar}) on the glucose substrate?

   c) What is the Y_{ATP} value?

2) If a fermenting organism growing on rich medium has a Y_{ATP} value of 9.3 g (biomass $mole^{-1}$ ATP) and Y_{molar} = 28 g (biomass $mole^{-1}$ substrate). What is the yield of ATP per mole of substrate utilised?

4.6 Growth rate and the cell cycle

asynchronous and synchronous cultures

The complete sequence of events extending from the formation of a new cell through the next division is called the cell cycle. So far in this chapter we have considered asynchronous growth of population, where, cells from all stages in the cell cycle are represented in the culture at the same time. Such cultures are not suitable for studies on the growth behaviour of individual cells. However, information about cell cycle events can be obtained by the study of synchronous cultures ie cultures composed of cells which are all at the same stage in the cell cycle. Measurements made on such cultures are equivalent to the measurements made on individual cells. In this section we will consider 1) how events in the cell cycle of individual cells are related to the growth rate and 2) how synchronous cultures can be obtained in the laboratory.

4.6.1 The bacterial cell cycle

The major cytological changes seen during the bacterial cell cycle are cell growth, division by septum formation and DNA segregation into daughter cells.

Most studies on cell cycle events in prokaryotes have used *Escherichia coli*; a rod shaped organism which, being prokaryotic, usually carries only one chromosome. A young *E. coli* cell growing at a constant rate will double in length without changing diameter, then divide into two cells of equal size by transverse fission. This involves extension of cell walls and membranes. For the cell wall, synthesis is restricted to particular regions of the pre-existing wall. Because each daughter cell receives at least one copy of the genetic material, DNA replication and cell division must be tightly coordinated.

DNA
replication is
coordinated
with cell
division

Mechanisms of DNA synthesis are considered elsewhere in the BIOTOL series, in this chapter we are only concerned with how DNA replication is coordinated with cell division. Important findings in *E. coli*, which have contributed to the understanding of cell cycle coordination, are:

- the period of DNA replication is more-or-less constant regardless of growth rate (or generation time);

- the time from end of a round of DNA replication to the formation of a division septum is constant at all except slow growth rates.

So, how is this timing achieved? The complete answer to this question is not known although current evidence indicates that two sequences of events, operating in parallel but independently, control division and the end of the cell cycle. Figure 4.6 illustrates the coordination of bacterial cell cycle events.

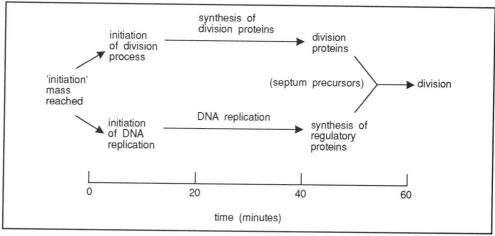

Figure 4.6 Events in the cell cycle in *E. coli*.

In Figure 4.6, a 60 minute generation time is assumed for simplicity; the actual generation time, of course, may vary. We can see that *E. coli* requires about 40 minutes to replicate its DNA and 20 minutes after termination of replication to prepare for division. Some important features of the replication cycle in bacteria are listed below:

- the cell cycle is triggered when the cell reaches a certain mass or volume;

- the control sequence for cell division and that for DNA replication involves the synthesis of cell cycle regulatory proteins;

- in the final stage the newly formed DNA copies are attached to adjacent sites on the plasma membrane. Membrane growth first separates them then a cross wall (or septum) forms between the two.

discontinuous
and continuous
DNA replication

Under optimum growth conditions the doubling time for *E. coli* is only about 20 minutes. However, we know that the DNA replication period alone is more-or-less constant at 40 minutes regardless of growth rate. So, how can we explain a short generation time in relation to what we know about cell cycle events? The explanation for this is based on the fact that the relationship of DNA synthesis to the cell cycle varies with the growth rate. When the doubling time is 60 minutes or longer DNA replication is a discontinuous process ie does not occur in the last 20 minutes of the cell cycle. However, when the culture is growing with a doubling time of less than 60 minutes, a second round of replication is initiated while the first round is still underway. The daughter cells may actually receive DNA with two or more replication forks, and replication is continuous because the cells are always copying their DNA. Figure 4.7 illustrates the replication of bacterial DNA at low and at high growth rates.

In principle, if the cells divide at 20 minute intervals then the initiation of DNA synthesis must occur at 20 minute intervals. (Think of trains going down a track: if they all travel at the same rate and they start at 20 minute intervals they will arrive at 20 minute intervals irrespective of how long the journey actually takes).

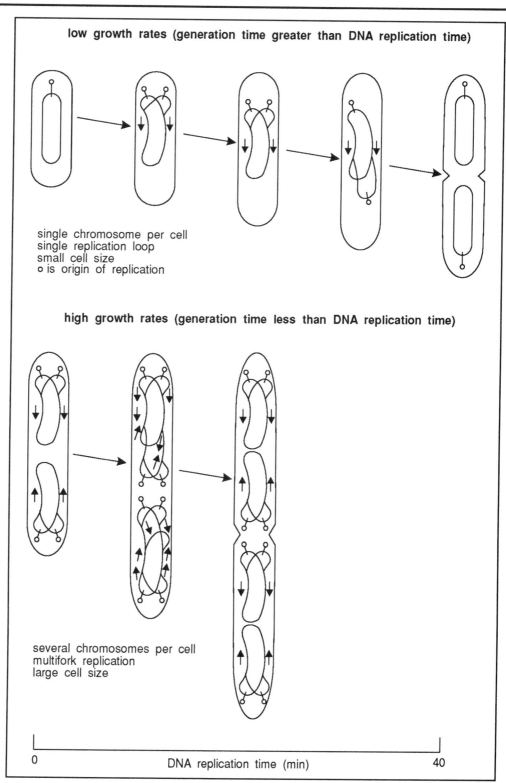

low growth rates (generation time greater than DNA replication time)

single chromosome per cell
single replication loop
small cell size
o is origin of replication

high growth rates (generation time less than DNA replication time)

several chromosomes per cell
multifork replication
large cell size

0   DNA replication time (min)   40

Figure 4.7 Replication of bacterial DNA at low and at high growth rates.

4.6.2 The eukaryotic cell cycle

A complete description of the eukaryotic cell cycle is beyond the scope of this chapter.

It is discussed in the BIOTOL text 'Infrastructure and Activities of Cells', 'In vitro Cultivation of Plant Cells' and 'In vitro Cultivation of Animal Cells'.

DNA replication in eukaryotes

In many respects the pattern of control of the cell cycle in eukaryotes resembles those found in bacteria with additional complexity due to the presence of organelles and the nuclear arrangement of DNA.

4.6.3 Cell synchronisation

In order to obtain enough material to study biochemical changes taking place during the cell cycle it is often necessary to synchronise the division of cells within a population. Since micro-organisms can adapt very rapidly to changes in their environment it is important that methods used to obtain population synchrony do not introduce artifacts. There are two general ways of obtaining synchrony:

induction and selection procedures

- by induction eg alternate temperature changes, light pulses to photosynthetic organisms or the use of metabolic inhibitors affecting essential functions in the cell cycle;

- by selection eg on the basis of size using density gradient centrifugation, or filtration, or by elution of newly divided cells from immobilised parents.

Two selection methods in current use are outlined in Figure 4.8.

∏ Can you think of an advantage that selection procedures have over induction procedures, for obtaining synchronised cultures.

Selection procedures are less likely to introduce artifacts since the cells are selected on the basis of physical characteristics (usually size) with the minimum of chemical disturbance.

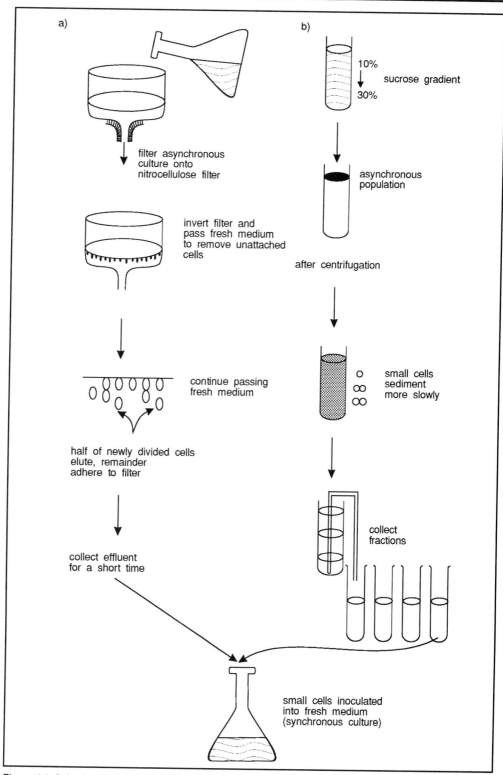

Figure 4.8 Selection methods for obtaining synchronous cultures. a) Membrane elution. b) Zonal centrifugation.

growth leads to
loss of
synchrony
    The growth of a synchronised culture of *E. coli* is shown in Figure 4.9. We can see that the number of cells in the culture remains approximately constant for about one hour while the newly formed cells grow in size. Then the number of cells doubles abruptly. In the second cycle, the plateau is less distinct and the population rise extends over a longer period, indicating that synchrony is already being lost. In the third division cycle, there is almost no indication of synchrony remaining.

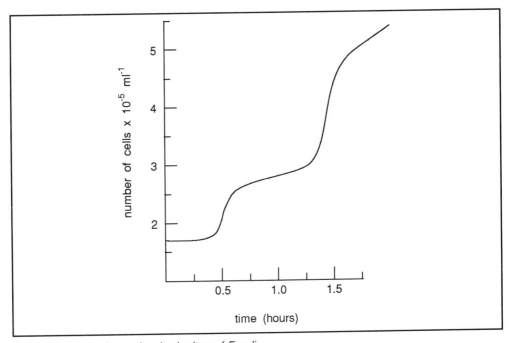

Figure 4.9 Growth of a synchronised culture of *E. coli.*

So, why does synchrony decay fairly rapidly on further growth? The explanation for this is that some cells within a population divide at an earlier age (smaller size or lower mass) than others. This may be related to the state of individual cells prior to the synchronisation procedure ie the history of individual cells. Figure 4.10 shows the distribution of generation times for exponentially growing *E. coli.*

∏   Describe how you would measure the degree of synchrony for a bacterial population.

This can be achieved in a variety of ways. For most bacterial cultures the simplest, and therefore the most common method, is to measure the period over which the population doubles. Here, a relatively short period reflects a relatively high degree of synchrony (see Figure 4.9).

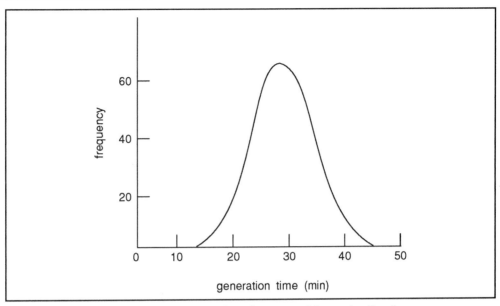

Figure 4.10 Theoretical distribution of doubling times for exponentially growing *E. coli.*

∏ We have seen that not all cells within a population divide at precisely the same age (time from the last division) or same size. How do you think cell size will influence the growth rate of cells? (Hint: examine Figure 4.10).

The relationship between cell size and growth rate is complex. The distribution of generation times in a population, shown in Figure 4.10, is a reflection of the cell size distribution. So, a plot of growth rate against cell size will generate a similar distribution: very small cells grow slowly, cells of intermediate size grow more rapidly, and very large cells again grow slowly.

continuous and discontinuous enzyme synthesis

It is easy to see that, theoretically at least, there should be twice as many young cells at age 0 as there are old cells of age t_d (mean cell age at division). In addition, direct microscopic observation has shown that individual cells of most micro-organisms increase in mass logarithmic with time and volume changes are also roughly logarithmically. So, what about the cell constituents, how do they change during the cell cycle? Usually there is a continuous increase of total protein and RNA during the bacterial cell cycle, but for individual enzymes the situation can be very different. If an enzyme were synthesised continuously throughout the cell cycle, in a synchronous culture, its activity would increase linearly with time until the gene coding for the enzyme doubled; then the rate should also double (assuming that the rate of enzyme synthesis is limited by the number of gene copies available for transcription). This is true for some enzymes but many appear to be synthesised discontinuously, each one appearing at a characteristic time which reflects the order on the chromosome of the genes coding for these enzymes.

∏ The next graph shows the theoretical change in cell number with time for a synchronous bacterial culture undergoing three rounds of division.

Draw on this graph the theoretical plots that you would expect for the following:

- change in cell mass of the population;

- change in total activity of an enzyme synthesised discontinuously in the middle of the cell cycle (Assume that the rate of enzyme synthesis is limited by the number of gene copies available for transcription);

- change in total activity of an enzyme synthesised continuously;

- change in cell volume of individual cells.

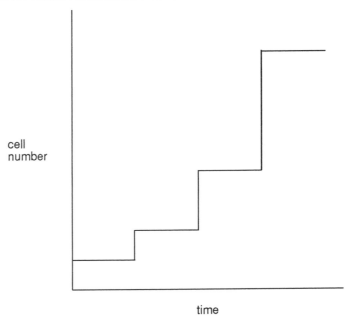

The correct response to this in-text activity is shown on the next page.

Π   If the rate of synthesis of a bacterial enzyme is limited by the number of gene copies, explain how you might expect the growth rate to influence the activity of the enzyme.

At high growth rates DNA replication becomes continuous and daughter cells may actually receive DNA with two or more replication forks (see Section 4.6.1). This will increase the number of copies of the gene coding for the enzyme. This may also increase the rate of enzyme synthesis and the activity of the enzyme.

Response to in-text activity.

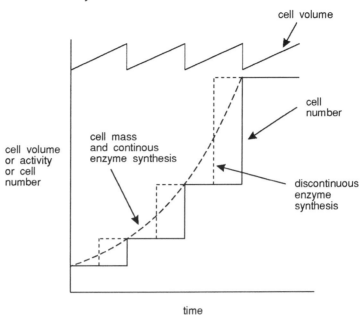

cell volume

cell
number

cell volume
or activity
or cell
number

cell mass
and continous
enzyme synthesis

discontinuous
enzyme
synthesis

time

SAQ 4.7

Identify each of the following statements as True or False, for the growth of *E. coli.*

1) The time taken to replicate DNA is largely independent of growth rate.

2) DNA replication is discontinuous at low growth rates.

3) At low growth rates there is no re-initiation at the origin of DNA replication and each cell contains one chromosome.

4) At high growth rates there is multiple initiation of replication at the origin and more than one chromosome per cell.

5) Daughter cells always receive exactly one copy of chromosomal DNA.

6) The mass of a cell is more-or-less constant at the time of cell division.

7) The mean cell cycle time is a measure of population synchrony.

Summary and objectives

In this chapter we have seen that in closed systems micro-organisms undergo a predictable pattern of growth, characterised by four main phases. After a period of adjustment (lag phase) the culture grows unrestricted at a rate close to μ_{max} (exponential phase). Changes occurring in the culture during exponential growth leads to a phase of low or zero growth rate (stationary phase). Further deterioration of culture conditions leads to a reduction in the number of viable cells (death phase). Growth during the exponential phase is characterised by the doubling time (t_d) and the specific growth rate (μ). According to the Monod relationship, μ is reduced at relatively low nutrient concentrations and the constant K_s reflects an organism's affinity for a nutrient. The maximum crop of biomass is also dependent upon limiting nutrient concentration; this is the principle underlying bioassays for nutrients. Whereas the growth yield per unit of limiting substrate (Y_s or Y_{molar}) varies amongst micro-organisms Y_{ATP} values are quite similar. This reflects the operation of different catabolic pathways in micro-organisms but the requirement for a constant amount of energy to polymerise macromolecules.

Growth in batch culture can be synchronised by induction or by selection methods. Studies on synchronised cultures are equivalent to studies on individual cells. Such studies have revealed that bacterial DNA replication is continuous at high growth rate but discontinuous at low growth rates.

Now that you have completed this chapter you should be able to:

- describe the various phases of growth in batch culture and factors influencing each phase;

- understand the mathematics of exponential growth;

- define and apply growth parameters (t_d, μ, k, Y_s, Y_{molar}, q, Y_{ATP});

- describe the Monod relationship and the meaning of K_s;

- outline methods for obtaining synchronous cultures and appreciate the advantages of these culture for cell cycle research;

- describe the prokaryotic cell cycle and the influence of growth rate on cell cycle events.

Growth in continuous culture

Growth in continuous culture

5.1 Introduction

Scientists are trained to conduct experiments in which only one variable is changed while all other variables are kept constant. This ideal situation is certainly not found in simple batch culturing of micro-organisms, which is the most commonly used technique in microbiology (Chapter 4). Once the medium is inoculated, several changes happen which are beyond control of the experimenter and which are due to exponential increase in cell number, decrease in nutrient concentration, change in pH and dissolved oxygen concentration, build up of toxic or inhibitory products.

In contrast, continuous culture methods enable constant cell numbers to be maintained in a constant chemical environment at specified growth rates for extended periods of time. A general set-up for continuous growth is shown in Figure 5.1.

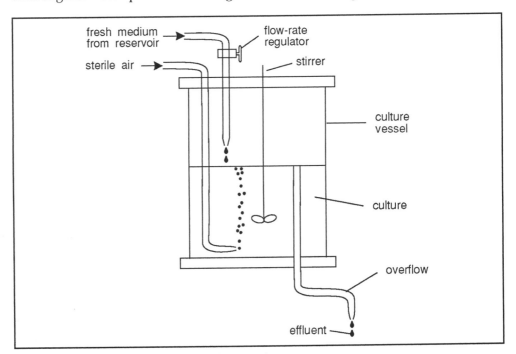

Figure 5.1 Set-up for continuous culture of micro-organisms.

In its simplest form the device consists of a culture vessel to which fresh medium with a growth limiting nutrient (often a carbon and/or energy source) is continuously added and spent medium, including cells, removed at the same rate. Such a device is called a chemostat. Cells cultured in a chemostat are all in the same physiological condition which is maintained throughout growth. This makes chemostat cultures ideally suited for (comparative) physiological studies on micro-organisms. Other applications can be

chemostat

found in the field of microbial ecology where continuous cultures can be used as model systems for studying the interaction between micro-organisms living in the same habitat.

In this chapter we shall focus on both theoretical and practical aspects of growth in 'flow-through' systems.

5.2 Continuous culture

So far we have only considered microbial growth in closed (batch) cultures (Chapter 4), in which growth takes place in a certain volume of suitable medium. In such a culture growth can only continue until the medium becomes unsuitable for further increase in biomass.

Initially the growth conditions will change only little, but once the increase in cell number, as a result of exponential growth, is well on the way considerable changes in growth conditions develop.

continuous cultures

If one wants to cultivate micro-organisms for a longer period in constant conditions batch cultures are not suitable and instead, so-called continuous cultures must be used.

A continuous culture consists of a culture vessel to which a continuous flow of fresh medium is added from a reservoir, at a constant rate. The medium's volume is kept at a constant level by an overflow, which allows surplus culture liquid (together with cells) to drain off (Figure 5.1).

chemostats control population density and growth rate

The most popular type of continuous culture is the chemostat in which both population density and growth rate of the organism can be controlled. So, how can the population density and the growth rate of the micro-organisms be regulated in a chemostat? To regulate the growth rate and the density of the micro-organisms it is necessary to vary the influx of nutrients per time unit. In a chemostat this is achieved by adapting the rate at which fresh medium is added and by adjusting the concentration of a growth-limiting nutrient.

A distinctive feature of a chemostat is that one of the nutrients (either the C, N, P, energy source or a growth factor) is present in the medium at such a low concentration that it becomes growth limiting. In Chapter 4 we saw that, if an essential nutrient is present at low concentration, the growth rate is proportional to that concentration. By regulating the rate at which fresh medium is added to the chemostat, we do, in fact, regulate the amount of limiting nutrient that is available for growth per time unit, and, thus a certain growth rate is forced upon the organism. Altering the flow rate leads to establishment (after a period of adjustment) of a new equilibrium which is characterised by a new (specific) growth rate and (possibly) a new cell density. The new dynamic equilibrium

steady-state

is called 'steady-state'. In this state the parameters by which the system is characterised do not alter anymore.

Another way of regulating the amount of limiting nutrient that is available for growth is to adjust its concentration in the fresh incoming medium. The effect of increasing the concentration is to establish, after a period of adjustment, a new steady-state at a higher cell density. The growth rate of the cells at the new steady-state is unaltered.

5.3 Mathematical relationships of growth in chemostats

growth rate
and specific
growth rate

In this section we will prove that, in a chemostat in steady-state, the specific growth rate (μ) of the culture equals the rate at which the culture is diluted. Before we go any further lets clarify the distinction between growth rate and specific growth rate.

We remind you of two relationships established in Chapter 4.

Growth rate = change in biomass concentration (x) with time = $\dfrac{dx}{dt}$

Specific growth rate (μ) = growth rate per unit amount of biomass=

$$\frac{1}{X} \cdot \left(\frac{dx}{dt}\right)$$

Note the convention we are using. Lower case x refers to the biomass concentration. Upper case X refers to the total biomass present in the vessel. It is important at this point that you make sure you understand the difference between growth rate and specific growth rate and that you learn the meanings of x and X.

∏   Write down some suitable units for growth rate and for specific growth rate.

Common units for growth rate are:
- cells h^{-1} or cells day^{-1};
- g biomass h^{-1} or g biomass day^{-1}.

Units for specific growth rate are always reciprocal of time:
- usually h^{-1} for bacterial cultures;
- usually $days^{-1}$ for plant and animal cell cultures.

In a chemostat the growth rate of the micro-organism is determined by the rate at which fresh medium is added to the growth vessel. As we have seen, this is caused by the limited availability of one of the necessary nutrients. What is the relationship between the specific growth rate (μ) and the rate of medium addition? To answer this question we need to go back to the moment the medium in the chemostat is inoculated. Figure 5.2 describes the development of growth in a chemostat in terms of cell density from the moment of inoculation until the steady-state is reached.

∏   Examine Figure 5.2 and describe, in your own words, the development of growth in a chemostat in terms of growth rate and availability of nutrients from the moment of inoculation until the steady-state is reached.

Now confirm your description with that given in the following paragraph.

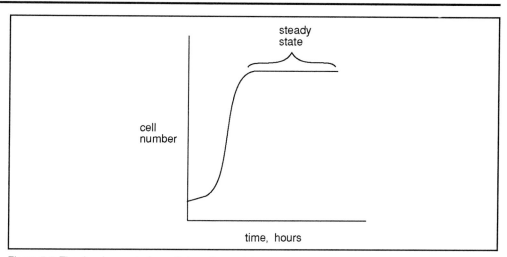

Figure 5.2 The development of growth in a chemostat.

After inoculation the culture is able to grow at maximal growth rate (μ_{max}) for some time, since all nutrients are, relative to the number of cells, in excess. Hence the population density increases with subsequent increased substrate consumption per time unit. Since one of the nutrients in the medium is present in limited supply, the increased need for substrate leads to a decrease in growth rate: the supply only partly covers the cells needs. As long as the growth rate exceeds the rate at which cell material leaves the overflow, the cell density in the vessel increases, and, consequently, the growth rate decreases. Decrease in growth rate continues till a steady-state is reached, in which the rate of increase in cell number just equals the rate at which cell material leaves the culture.

The relationship between growth rate or specific growth rate and medium flow can be described mathematically. The medium flow through the system is then represented by the term dilution rate, D, which is defined as:

dilution rate

$$D = \frac{F}{V}$$

(E - 5.1)

in which:

D = dilution rate (h^{-1})
V = culture volume (l)
F = flow rate ($l\ h^{-1}$)

The flow rate can be regulated by using a pump. The rate of change in the amount of biomass in the vessel is given by the increase in biomass by growth - amount lost via the overflow.

Thus:

| rate of change in biomass | = | increase in biomass by growth | - | loss of biomass via the overflow |

Therefore: $\dfrac{dx}{dt} = \mu X - Fx$ (E - 5.2a)

(remember that X = total amount of biomass and x = biomass concentration, μ = specific growth rate, F = flow rate.

If we divide through by the volume (V) of the vessel then Equation 5.2a becomes:

$$\dfrac{dx}{dt} = \mu x - \dfrac{F}{V} x = \mu x - Dx$$ (E - 5.2b)

If the specific growth rate is greater than the dilution rate ($\mu > D$), as is the case at the start of the culture, the biomass concentration will increase. This will continue until the steady-state is reached: the number of cells leaving the vessel equals the number of cells that develop by growth. In that situation:

$$\dfrac{dx}{dt} = 0, \text{ or } \mu x = Dx$$

dilution rate controls specific growth rate

So the steady-state is reached when $\mu = D$. A continuous culture in its steady-state can be considered as a self-regulating system. For instance, if the growth rate becomes smaller than the dilution rate ($\mu < D$) the biomass concentration temporarily decreases (a negative biomass balance exists), because of which the substrate concentration in the vessel increases. This, in turn, causes μ to increase again until a new balance is reached ($\mu = D$).

In practice it is usually assumed that the steady-state is reached when the biomass concentration has not changed during two volume changes and at least a total of five volume replacements have taken place since a new dilution rate was set.

∏ Complete the table using the words 1) Constant, 2) Variable, 3) Increasing, 4) Decreasing.

	Exponential phase of batch culture	Chemostat operating in steady-state
Growth rate of culture		
Specific growth rate of culture		
Biomass		
Available nutrients		
Culture volume		
Toxic metabolites		

For exponential phase of batch culture the correct sequence of responses is 3, 1, 3, 4, 1, 3. For a chemostat operating in steady-state response 1 (constant) is appropriate in all cases. You will note that in exponential phase the growth rate of the batch culture increases, this is caused by the increase in biomass ie as biomass increases there are more cells (catalysts) contributing to the increase of biomass per unit time (growth rate). However, the doubling time of cells in this phase is constant.

the Monod relationship

The mechanism on which the regulating effect of the dilution rate on the growth rate is based is the relationship between μ and the concentration of the growth-limiting nutrient. In the culture vessel this concentration is nearly always so low that $\mu < \mu$max. The empirically derived equation for the relationship between specific growth rate and substrate concentration is formulated by Monod as follows:

$$\mu = \mu_{max} \frac{s}{K_s + s}$$

(E - 5.3)

in which:
μ_{max} = the maximally possible growth rate (at saturating substrate concentration)
s = substrate concentration in the culture liquid
K_s = Monod constant (numerically equivalent to the substrate concentration, with $\mu = 0.5\ \mu_{max}$, also known as the saturation constant).

As we showed earlier, in steady-state μ = D, which changes Equation 5.3 to:

$$D = \mu_{max} \frac{\bar{s}}{K_s + \bar{s}}$$

(E - 5.4)

in which:
$\bar{s}$ = steady-state substrate concentration in the culture liquid. Note our nomenclature, the bar above the s signifies the steady state. Rearranging this equation leads to:

$$\bar{s} = K_s \frac{D}{\mu_{max} - D}$$

(E - 5.5)

This equation demonstrates how the steady-state substrate concentration in the chemostat is determined by the dilution rate.

∏ Use Equations 5.4 to show in which conditions it is possible for the specific growth rate in a chemostat to be about the same as μ_{max}.

If the steady-state concentration ($\bar{s}$) is much greater than K_s (for example $\bar{s} > 10\ K_s$) then, in Equation 5.4, μ_{max} is multiplied by a figure close to unity (one) ie $D \approx \mu_{max}$. We know that a steady-state μ = D, therefore $\mu \approx \mu_{max}$. We can see from Equation 5.5 that $\bar{s}$ can only be far greater than K_s at high dilution rates.

∏ A chemostat operating in steady-state at a dilution rate of 0.1 h^{-1} sets a limiting nutrient concentration of 0.5 micro moles 1^{-1}. Determine the Monod constant (in suitable units) if μ_{max} for the organism is 0.5 h^{-1}.

We can use Equation 5.5 to solve this. Rearranging this equation gives:

$$K_s = \bar{s}\,\frac{\mu_{max} - D}{D}$$

$$K_s = 0.5\,\frac{(0.5 - 0.1)}{0.1}$$

$$K_s = 0.5 \quad . \quad 4$$

$$K_s = 2 \text{ micro moles } l^{-1}$$

affinity for limiting substrate

Figure 5.3 illustrates the meaning of K_s by showing the μ against substrate concentration (s) curves for growth of two micro-organisms, A and B, on the same substrate. The maximum growth rate of both organisms is the same but K_s of organism A is lower than that of organism B. This means that at low substrate concentrations A's affinity for the substrate is higher than that of B, which is expressed by a higher specific growth rate. So at these low substrate concentrations A is in a favourable position when competing with B.

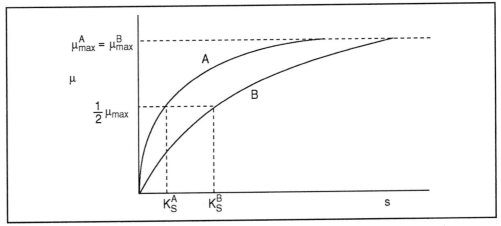

Figure 5.3 Monod kinetics for growth of two different micro-organisms (A and B) on the same substrate. μ = specific growth rate; s = substrate concentration; K_s = Monod constant.

changes in substrate concentration

The changes occurring in the substrate concentration of the culture liquid in a chemostat, before a steady-state is achieved can be described by the following equation:

$$\frac{ds}{dt} = \text{substrate input - substrate used for growth - substrate output}$$

or, the change in substrate concentration is equal to the substrate input reduced by the amount of substrate used for growth together with the amount of substrate leaving the vessel via the overflow.

This can be expressed as follows:

$$\frac{ds}{dt} = DS_R - \frac{\mu x}{Y} - Ds \qquad\qquad (E - 5.6)$$

in which:
S_R = substrate concentration in the fresh medium reservoir.
Y = microbial yield on growth limiting substrate.
s = substrate concentration in the chemostat.

The expression for growth in this equation $(\frac{\mu x}{Y})$ signifies the substrate consumption when μx biomass is formed.

It is derived as follows:

Substrate used for growth = growth rate/yield (where $Y = \dfrac{dx}{ds}$)

Thus:

$$\left(\frac{ds}{dt}\right)_G = \frac{1}{Y} \cdot \left(\frac{dx}{dt}\right)$$

Note we have used subscript G to denote substrate used for growth.

We already know that $\mu = \dfrac{1}{X} \cdot \left(\dfrac{dx}{dt}\right)$ or $\dfrac{1}{x} \cdot \left(\dfrac{dx}{dt}\right)$

So: $\dfrac{dx}{dt} = \mu x$

Then: $\left(\dfrac{ds}{dt}\right)_G = \dfrac{\mu x}{Y}$

As long as the steady-state is not reached the values of the last two terms of Equation 5.6 keep changing. Only when the steady-state is reached at $\dfrac{ds}{dt} = 0$ we find:

$$\frac{ds}{dt} = 0 = DS_R - \frac{\mu \bar{x}}{Y} - D\bar{s} \tag{E - 5.7}$$

in which $\bar{x}$ = biomass concentration in the steady-state and $\bar{s}$ is the substrate concentration in steady-state. Equation 5.7 can be simplified ($\mu = D$) to:

$$\bar{x} = Y (S_R - \bar{s}) \tag{E - 5.8}$$

If you think about this equation for a moment you will realise that it makes sense. Since S_R is the substrate concentration of the input and $\bar{s}$ is the steady-state substrate concentration, then $S_R - \bar{s}$ represents the amount of substrate used for growth. If we multiply this by the growth yield (Y), then this tells us how much biomass is produced thus $\bar{x} = Y (S_R - \bar{s})$.

After combination of Equations 5.8 and 5.5 $\bar{x}$ becomes:

$$\bar{x} = Y \left(S_R - K_s \frac{D}{\mu_{max} - D} \right)$$

(E - 5.9)

The parameters in the equations already mentioned, K_s, μ_{max} and Y, represent values for a particular micro-organism growing under a certain set of conditions (temperature, composition of growth medium, etc), whereas S_R and D are the variables that can be regulated by the experimenter. When S_R is increased $\bar{x}$ increases as well (Equation 5.8), leaving $\bar{s}$ unchanged (Equation 5.5). In addition remember that μ will increase when D is increased ($\mu = D$), which means that $\bar{s}$ must increase as well (Equations 5.4 and 5.5). In the latter case the biomass concentration, $\bar{x}$, decreases (Equation 5.8) as less substrate will have been used for conversion into biomass.

If, for a certain micro-organisms K_s, μ_{max} and Y, are known we can predict the micro-organisms behaviour in the chemostat, using Equations 5.5 and 5.9. Figure 5.4 shows the dependence of biomass concentration ($\bar{x}$) and substrate concentration ($\bar{s}$) on the dilution rate (D) and the substrate concentration in the reservoir (S_R).

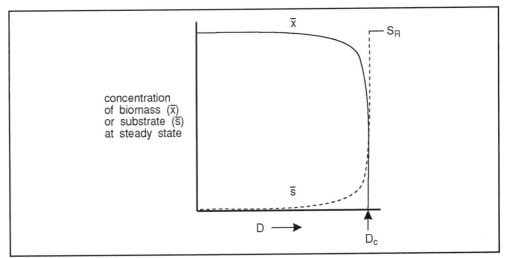

Figure 5.4 Relationship between biomass concentration ($\bar{x}$) and substrate concentration ($\bar{s}$) plotted against dilution rate (D). S_R = substrate concentration in reservoir. D_c = critical dilution rate. Remember D = μ at steady-state.

the critical dilution rate

There is, of course, a limit to which the dilution rate can be increased. From a certain value of D onwards the output of biomass via the overflow will be greater than biomass production by growth. At that stage the culture can no longer establish a steady-state and is drained from the chemostat. The critical value for D (D_c) is reached when the culture just washes out. Under these conditions, $\bar{x}$ = 0 and $\bar{s}$ = S_R. The value of D_c follows from the equation:

$$D_c = \mu_{max} \frac{S_R}{K_s + S_R}$$

(E - 5.10)

In practice, when $S_R > K_s$ (which is often the case) D_c approaches μ_{max}. A steady-state can be reached with growth rate close to μ_{max}. However, if S_R is very low, the culture will be washed out at $\mu << \mu_{max}$.

Π Figure 5.4 shows that with increasing D, $\bar{s}$ increases only slightly over a wide range of D values. Only at much greater values of D does $\bar{s}$ show sizable increases. Can you give an explanation for this effect?

We can explain the relationship shown in Figure 5.4 in several ways. If you look back at Equation 5.4 and 5.5 you will see that the steady-state substrate concentration $\bar{s}$ is related to K_s and D by the following relationship.

$$\bar{s} = K_s \frac{D}{\mu_{max} - D}$$

When D is small, relative to μ_{max} then the value of $\dfrac{D}{\mu_{max} - D}$ is also small and the value of $\bar{s}$ will be small. As D increases then the value of $\dfrac{D}{\mu_{max} - D}$ also increases and thus $\bar{s}$ must also increase. This effect becomes even more pronounced as D approaches μ_{max}.

Another way of thinking about this is to realise that if the organism has a low K_s value (ie a high affinity for its substrate) then at low dilution rates, the cells have sufficient time to consume virtually all of the substrate and $\bar{s}$ is very low. As the dilution rate is increased, the residence time of the media in the vessel is short, the cells have insufficient time to use all (or most) of the substrate, thus $\bar{s}$ rises.

A third way to explain the relationship shown in Figure 5.4 is to use the Monod relationship directly. Since $\mu = \dfrac{s}{K_s + s}$ and at steady-state $D = \mu$ then to maintain a high value of μ then we must have a high concentration of substrate. Thus at high dilution rate $\cong$ high values of μ), then $\bar{s}$ must also be high.

Π Some organisms display a very poor affinity for substrate (they have a high K_s value). In these cases, what effect does an increase in D have on the substrate ($\bar{s}$) and biomass ($\bar{x}$) concentration in the steady-state?

In these cases a relatively small increase in D results in considerable adjustment (increase) of $\bar{s}$. The effect is illustrated in Figure 5.5; compare this with Figure 5.4 which is typical for low K_s values.

productivity The amount of cell material produced per time unit is known as productivity. In a chemostat productivity will then be $D\bar{x}$.

Π Examine Figure 5.4 and try to predict how productivity would change with increasing D.

Figure 5.4 shows that productivity should rise to a maximum with increasing D, and then decline sharply because of strong decrease in $\bar{x}$.

In Chapter 4 we saw that the relationship between generation time (t_d) and growth rate was: $\mu = \ln2/t_d$. Since in the steady-state $\mu = D$, t_d can be calculated from $D = \ln2/t_d$.

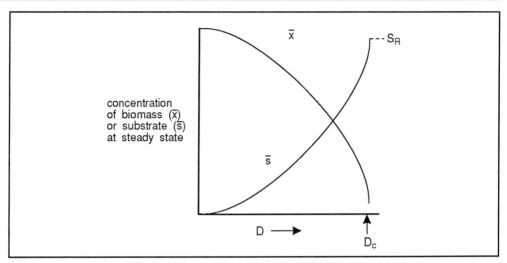

Figure 5.5 Relationship between $\bar{x}$ and $\bar{s}$ plotted against D for an organism with a relatively high K_s value for substrate; compare this with Figure 5.4.

Π A bacterium is cultivated in a chemostat with glucose as growth limiting nutrient. The growth constants have the following values: μ_{max} = 1 h^{-1}, K_s = 0.1g l^{-1}, and Y = 0.5g g^{-1}. Demonstrate in a graph the relationship between $\bar{x}$ and $\bar{s}$ as a function of D, presuming S_R = 10g l^{-1}.

Also, on the same graph, draw the relationships for productivity as a function of D.

(When calculating values for $\bar{x}$, $\bar{s}$, and productivity, use the following dilution rates (h^{-1}): 0.2, 0.4, 0.8, 0.95 and 0.98).

Use a table like that below to record your calculated values of $\bar{x}$, $\bar{s}$.

D	$\bar{s}$	$\bar{x}$	Productivity
0.2			
0.4			
0.6			
0.95			
0.98			
1.00			

Figure 5.6 shows the correct plots for the organism. Compare your graph with this figure.

You should have obtained for $\bar{s}$ using Equation 5.5 and values for $\bar{x}$ using Equation 5.9. Productivity = D $\bar{x}$.

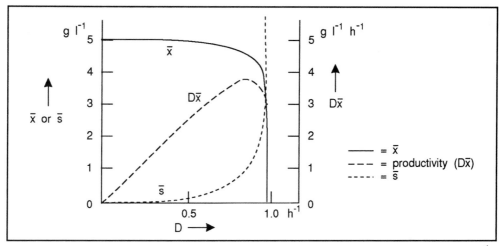

Figure 5.6 Growth behaviour of an organism characterised by the values of growth constants: $\mu_{max} = 1\ h^{-1}$; $K_s = 0.1g\ l^{-1}$; $y = 0.5g\ g^{-1}$.

Here are the results of our calculations.

D (h^{-1})	$\bar{s}$ (g l^{-1})	$\bar{x}$ (g l^{-1})	productivity (g l^{-1} h^{-1})
0.2	0.025	4.99	0.998
0.4	0.06	4.97	1.988
0.6	0.15	4.72	2.952
0.95	1.9	4.05	3.85
0.98	4.9	2.55	2.5
1.00	$\bar{s} = S_R$	0	0

Note that the substrate concentration when $D = \mu_{max}$ is equal to S_R as we have wash out of the culture.

Note also how the productivity rises as D increases until this is counteracted by a decrease in $\bar{x}$.

1) You are growing two micro-organisms (A and B) separately in a chemostat, with 0.2% glucose as limiting substrate. Both organisms have the same maximum growth rate (0.5 h^{-1}) but a different K_s for glucose (10^4 and 10^{-2} mol l^{-1} respectively). At which values for D will these organisms wash out? (Molecular mass of glucose = 180).

2) A bacterium, known to have a μ_{max} of 2 h^{-1}, is grown in a chemostat with a working volume of 2 l^{-1}. Fresh medium is fed to the culture at a rate of 1.0 l h^{-1} and the concentration of limiting substrate in this medium (S_R) is 5g l^{-1}. In the chemostat the steady state concentration of biomass ($\bar{x}$) is 2g l^{-1} and of limiting substrate ($\bar{s}$) is 0.05g l^{-1}. Use suitable units and calculate:

   a) the specific growth rate (μ) in steady state;

   b) the Monod constant (K_s) for the limiting substrate;

   c) the growth yield (Y) on limiting substrate.

5.4 Determination of growth constants and growth yield

determination of μ_{max} and Ks

The value of the growth constants μ_{max} and K_s can be determined in several ways. One possibility is to measure $\bar{s}$ for various values of D. With the aid of Equation 5.5, μ_{max} and K_s can then be calculated. However, in practice $\bar{s}$ can only be measured for a short range of D values (only at high dilution rates can differences in $\bar{s}$ be measured accurately, see Figure 5.4) which results in inaccurate values for μ_{max} and K_s. Because of this, Equation 5.4 is usually converted to an equation for a straight line (generalised form, y = mx + c) by taking reciprocals:

$$\frac{1}{D} = \frac{1}{\mu_{max}} + \frac{K_s}{\mu_{max}} \cdot \frac{1}{\bar{s}}$$

(E - 5.11)

Plotting $\dfrac{1}{\mu}$ against $\dfrac{1}{\bar{s}}$ results in a straight line which intercepts the y-axis at $\dfrac{1}{\mu_{max}}$ and the x-axis at $-\dfrac{1}{K_s}$, with K_s/μ_{max} as slope. Table 5.1 shows a number of examples of experimentally determined values for K_s for a variety of nutrients.

Organism	Nutrient	K_s (mg l^{-1})
Escherichia coli	Glucose	6.8×10^{-2}
Escherichia coli	Phosphate ions	1.6
Pseudomonas sp.	Methanol	0.7
Aspergillus niger	Glucose	5.0
Saccharomyces cerevisiae	Glucose	25

Table 5.1 Experimentally determined values for K_s. Note that if we convert K_s into mol l^{-1}, values typically fall in the range 10^{-5} - 10^{-7} mol l^{-1}.

Π   Use the data shown below to determine, as accurately as possible, values for the growth constants μ_{max} and K_s using Equation 5.11.

D (h^{-1})	$\bar{s}$ (mg l^{-1})
0.30	5
0.35	7
0.40	10
0.50	25

Attempt this before reading on.

You should have plotted the reciprocals of D and $\bar{s}$.

Based on these data a suitable plot, for determination of growth constants is shown in Figure 5.7. Values of K_s and μ_{max} are given in Figure 5.7.

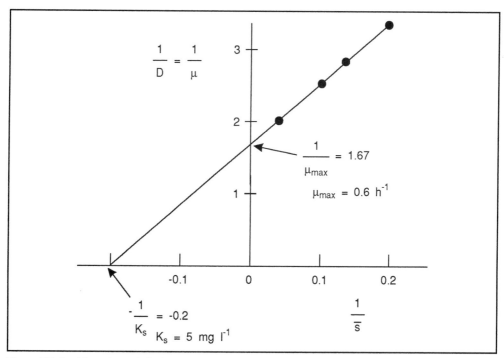

Figure 5.7 Double reciprocal plot for the determination of growth constants (response to in-text activity).

In batch cultures the cell yield values (Y) can be obtained by producing a series of cultures with increasing substrate concentration. After growth has stopped the remainder of the substrate, and the biomass formed can be measured. When the amount of substrate used is plotted against the increase in biomass, the slope of the resulting straight line represents the yield.

To determine the yield in chemostat cultures we should be able to use Equation 5.8. If we could measure $\bar{x}$ and $\bar{s}$ accurately Y would follow from this equation. However, when doing so, the yield proves to deviate significantly from values found in batch cultures and is strongly dependent on D. The deviations are greatest at low D and occur especially when the energy source is the growth limiting substrate. This can be explained by assuming that per time unit, independent of μ, a cell needs a certain amount of energy for maintenance. This maintenance energy is, for instance, needed for maintaining the proton motive force which is, among other purposes, used for osmotic activity (maintaining the ion gradients across the cell membrane). Furthermore, energy is needed for the turnover of proteins and mRNA, for repair, and for movement. That micro-organisms require maintenance energy can be deduced from the fact that many of them show a certain degree of respiration during periods of rest (no growth). Assuming the cell needs a certain constant amount of maintenance energy per time unit, the percentage of the total amount of available substrate which is used for maintenance increases with decreasing D (=μ). As a consequence, the measured yield (Y) at low dilution rate is lower than theoretically expected.

maintenance energy influences Y

∏ What effect on the relationship between steady-state biomass concentration and dilution rate would you expect 1) a relatively low maintenance energy value and 2) a relatively high maintenance energy value, to have? Show the effects in a graph ($\bar{x}$ versus D). Try to do this before looking at our response.

The effects are shown in Figure 5.8.

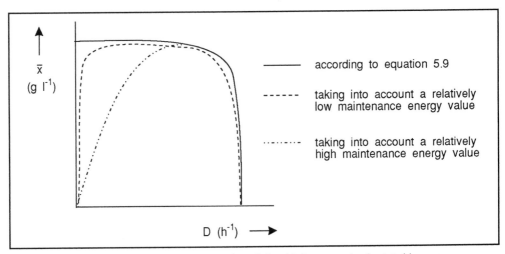

Figure 5.8 The effect of maintenance energy on the relationship between steady-state biomass concentration and dilution rate (response to in-text activity).

Pirt constructed a formula in which the experimentally determined yield (Y) is related to the calculated yield (Y_G) based on the amount of substrate used for growth. The starting point is the following equation:

overall rate   =   rate of substrate   +   rate of substrate
of substrate utilisation   utilisation for growth (G)   utilisation for maintenance (M)

or: $\dfrac{ds}{dt} = \left(\dfrac{ds}{dt}\right)_G + \left(\dfrac{ds}{dt}\right)_M$

(E - 5.12)

When expressing the overall substrate utilisation and the substrate utilisation for growth in yield terms, we find,

since, as we saw for Equation 5.6, $\left(\dfrac{ds}{dt}\right)_G = \dfrac{\mu x}{Y}$

$$\dfrac{\mu x}{Y} = \dfrac{\mu x}{Y_G} + mx \qquad\qquad\qquad\qquad \text{(E - 5.13)}$$

in which:
Y = g biomass g^{-1} substrate used
Y_G = g biomass g^{-1} substrate used for growth
m = maintenance coefficient (g substrate used for maintenance per g biomass per hour).
Rearrangement of (5.13) produces the following relationship:

$$\dfrac{1}{Y} = \dfrac{1}{Y_G} + \dfrac{m}{\mu} \text{ or } \dfrac{1}{Y} = \dfrac{1}{Y_G} + \dfrac{m}{D} \qquad\qquad \text{(E - 5.14)}$$

determination of m

When 1/Y is plotted against 1/D we find the value for m (slope) and 1/Y_G (interception on the Y axis) graphically. Quite often, when experimental data are plotted we find indeed a linear relationship. In these cases it is presumed that m is constant over the whole range of possible growth rates. Typical values, based on mass of substrate and cells are 0.02 to 0.04g g^{-1} h^{-1}. In those cases where no linear relationship is found m is obviously dependent on the growth rate.

SAQ 5.2	Explain how you would determine a value for the maintenance energy (m) and the calculated yield (Y_G), using a chemostat and methods for measuring biomass and substrate concentrations.

5.5 Turbidostat

The flow-controlled chemostat, in which the pump supplying the medium determines the dilution rate, is generally used at relatively low dilution rates, but is unsuitable for growth rates close to the critical dilution rate, $D_C \approx \mu_{max}$.

turbidostats control biomass concentration by altering flow rate

If one wants to cultivate micro-organisms in continuous culture at a growth rate in the region of μ_{max}, a turbidostat is mostly used. In this device it is not the pumping rate that is fixed but the biomass concentration. To establish this, in a turbidostat the density (turbidity) of the culture is monitored and kept constant by an electronic proportional adjustment of the pumping rate.

Π   Examine Figure 5.4 and explain, in your own words, why chemostat operation of a continuous culture is generally performed over a different range of dilution rates to that used for turbidostat operation.

The figure shows that, when μ is approaching μ_{max}, a small change in pumping rate (ie in D) causes a dramatic effect on the steady-state biomass concentration, $\bar{x}$. Deviations in pumping rate can even cause the culture to wash out. In a chemostat, maximum stability (ie maintaining the established steady-state) is obtained in the region where a slight change in D only causes a minimal change in biomass concentration.

For a turbidostat, the biomass concentration controls the flow rate (and therefore the dilution rate). Such a control functions optimally if a small change in flow rates produce a relatively large change in biomass concentration. Maximum stability of a steady-state in a turbidostat is therefore achieved at high dilution rates ie over the region of sharply declining $\bar{x}$ in Figure 5.4.

A turbidostat is particularly useful for research on eukaryotic cells. Their μ_{max} is so much lower than that of bacterial cells that it is impractical to grow them in chemostats. Continuous culture of these cells must be carried out at μ values approaching μ_{max}. A complicating factor, when using a turbidostat, is the fact that micro-organisms tend to develop growth on the actual optical sensor which then becomes covered by a biofilm.

If there is a biofilm on the sensor's surface this is read as cell density and the flow rate will be increased accordingly. This will lead to a wash out of the culture (μ is close to μ_{max}). One way of avoiding this problem is to keep the flow rate through the sensor high, so that cells can not adhere to and grow on it. Another approach is the use of a sensor that can be removed for cleaning and resterilisation, and then can be returned to the vessel.

∏ Can you think of any other circumstances when a turbidostat, measuring turbidity using an optical sensor, cannot be used.

If the turbidity which is the result of microbial growth cannot be measured. This is for instance the case when other particles apart from micro-organisms occur in the culture liquid. Another possibility is that the measured turbidity is not representative of the number of cells. This arises when the micro-organisms are not unicellular (for instance thread forming bacteria or filamentous fungi).

5.6 Fed-batch cultures

fed-batch The term fed-batch culture is used for batch cultures which are fed, continuously or intermittently with fresh medium without removal of culture fluid. Thus, the volume of a fed-batch culture increases with time.

The kinetics of a fed-batch culture can be described as follows. If growth of an organism in a batch culture is limited by the concentration of one of the nutrients, the biomass concentration at the end of the exponential phase (x_{max}) equals:

$x_{max} = Y.S_R$ (compare this equation with Equation 5.8).

The equation $x_{max} = Y.S_R$ assumes that the concentration of biomass in the inoculum (x_o) is negligibly small and that all the substrate in the medium is used up after growth ($\bar{s} = 0$).

∏ Try to write an equation which would better represent the biomass concentration at the end of the exponential phase.

The correct response is:

$$x_{max} = x_o + Y (S_R - s)$$

where:
x_o is the inoculum concentration
s = the substrate remaining in the culture

If fresh medium is added to the culture at a dilution rate lower than μ_{max} all (limiting) substrate is utilised as it is added to the system. Thus we can write:

$$F \cdot S_R = \mu \cdot \frac{X}{Y}$$

in which:
F = flow rate
X = total amount of biomass in the culture vessel (which equals the product of volume V and biomass concentration x).

In the culture vessel X will increase with time, with x remaining virtually constant:

$$\frac{dx}{dt} \cong 0$$

quasi
steady-state

So one might, in fact, say that in this case the system has, in a sense, reached a steady-state (sometimes called quasi steady-state), in which $\mu = D$.

∏ Try to describe the essential difference between the steady-state in a chemostat and the quasi steady-state in a fed-batch culture.

In a chemostat the dilution rate ($D = F/V$), and with it the specific growth rate (μ) are constant in the steady-state. In a fed-batch culture D changes (and hence μ) in time since the culture volume continues to increase.

As time progresses and the culture volume increases, the dilution rate (and therefore μ) will decrease. The value for D is given by the expression:

$$D = \frac{F}{V_0 + F \cdot t}$$

in which:
V_o = initial volume of the culture
t = period of time in which fed-batch conditions have been operating.

The dilution rate in a fed-batch culture can be kept constant by exponentially increasing the flow rate, using a computer controlled system.

applications

Development of fed-batch cultivation has been particularly stimulated by its application in industrial production of microbial metabolites and biomass, where it has certain advantages over ordinary batch techniques. One example is the production of baker's yeast. It is known that relatively low concentrations of certain sugars (saccharose, glucose), repress respiration (the so-called glucose-effect) and the yeast cells switch to fermentative metabolism, even in aerobic conditions. This, of course, has a negative effect on its yield. When maximum biomass production is aimed at, fed-batch techniques are the best choice, since then the concentration of the limiting nutrient, sugar, remains low enough to avoid repression of respiration.

SAQ 5.3

Identify each of the following statements as True or False.

1) Biomass productivity of a chemostat increases with increasing dilution rate up to the critical dilution rate (D_c).

2) In a plot of $\frac{1}{\mu}$ against $\frac{1}{s}$ the slope = $-\frac{1}{K_s}$.

3) In a plot of $\frac{1}{Y}$ against $\frac{1}{D}$ the slope of a straight line = m (maintenance coefficient).

4) In a chemostat the residual limiting nutrient concentration (s) decreases as μ approaches μ_{max}.

5) In a chemostat increasing the concentration of the limiting nutrient in the incoming medium (S_R) will establish a new steady state at a higher specific growth rate (μ).

6) In a chemostat the specific growth rate equals the flow rate.

7) In a chemostat the maintenance coefficient influences biomass yield most markedly at high dilution rates.

8) A relatively high K_s value reflects low affinity for growth substrate.

9) Turbidostats operate at dilution rates close to the critical dilution rate.

10) Steady state limiting nutrient concentrations (s) are generally higher in a turbidostat than in a chemostat.

11) In fed-batch culture $D = \frac{F}{V}$.

12) The maintenance coefficient is always independent of growth rate.

5.7 Applications of the different types of microbial cultivation techniques

5.7.1 Batch and fed-batch cultures

Batch cultures are often used on a laboratory scale to obtain cell material with which (biochemical) research can be carried out. Also, if one wants to have cells available which are in a certain phase of growth (for instance exponential growth) batch cultures are the answer. Some properties for instance are related to a particular phase in the growth cycle; for example, transformation of bacterial cells (uptake of foreign DNA) mainly takes place in the exponential phase. Batch cultivation is often carried out in so-called fermentors (Figure 5.5). Basically this is a large vessel containing culture liquid and which is usually provided with some peripheral equipment for optimal mixing and aeration, and temperature and pH control. Details of fermentor operation are given in Chapter 7.

fermentors

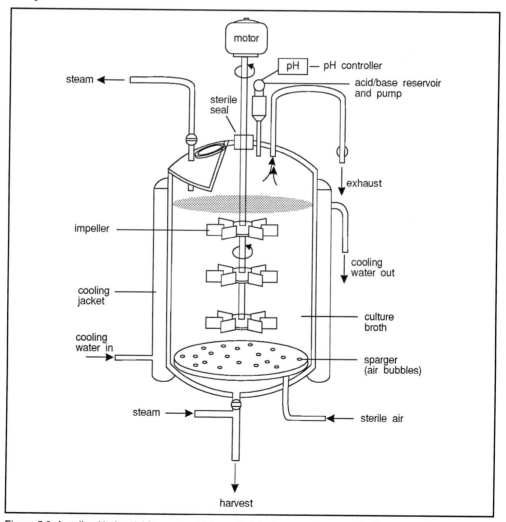

Figure 5.9 A stylised industrial fermentor (Note control devices are discussed in Chapter 7). Adapted from Brock, T.D and Madigan, J.L (1988) Biology of Micro-organisms, Prentice Hall.

⊓⊓ Why is the fermentor shown in Figure 5.9 not a batch culture in the strict sense of
     the word?

According to its definition a batch culture is a closed system to which during growth
nothing is either added or drained off. Yet, this happens in a fermentor, for instance
when a neutralising base is added where acid products are formed.

Fermentors can vary greatly in volume, from 2-10 l for laboratory purposes, to 400,000 l
for use in industry. Table 5.2 provides a survey of fermentors of various sizes and the
processes they are used for.

Size of fermentor, litres	Product
1-20 000	Diagnostic enzymes, substances for molecular biology
40-80 000	Some enzymes, antibiotics
100-150 000	Penicillins, aminoglycoside antibiotics, proteases, amylases, steroid transformations, amino acids
up to 450 000	Amino acids (glutamic acid)

Table 5.2 Fermentor sizes for various processes. Data derived from Brock, T.D and Madigan, J.L (1988)
Biology of Micro-organisms, Prentice Hall.

The development of fed-batch cultures has received a great boost because of their use
for the industrial production of biomass and metabolites in batch processes. We have
already touched upon the production of baker's yeast as one of the earliest examples.
Although continuous culture allows the highest degree of control over growth and
physiology of cells, its use in the fermentation industry is extremely limited, as we shall
see in Section 5.7.2. Fed-batch cultures have provided the fermentation technologist
with a valuable tool for controlling the environment of the fermentation process. They
can extend the productive period of a traditional batch culture without the inherent
disadvantage of a continuous process. In the most common type of fed-batch system the
fermentation is fed by one component of the medium (usually a carbon source), while
the feed rate is controlled by some measurable parameter of the fermentation process,
such as dissolved oxygen or pH.

The major advantage of feeding a culture with a medium component, rather than
incorporating it entirely in the initial batch, is that the nutrient concentration can be kept
low throughout the process.

| SAQ 5.4 | A constant low nutrient level is favourable for which of the following? |

1) Growing micro-organisms at maximal growth rates.

2) Avoiding possible toxic effects of the limiting nutrient.

3) Maintaining conditions in the culture within aeration capacity of the fermentor.

4) Removing repressive effects of the limiting nutrient.

5.7.2 Continuous cultures

research on microbial physiology

Continuous cultures (chemostat, turbidostat) are virtually exclusively used for scientific research. Application for industrial purposes is limited though in the following section we shall discuss the various applications using some examples. The chemostat provides us with a unique instrument to study the physiology of micro-organisms in controlled conditions at different growth rates and with different growth limiting nutrients. Such research has made clear that the composition of the cell, both qualitative and quantitive, changes greatly along with the growth rate. Additionally, much information has been obtained about regulating mechanisms of many metabolic processes. Let us examine some examples.

glutamate dehydrogenase

glutamine synthetase

Many bacteria have two pathways for ammonia assimilation and which one functions depends on the external ammonia concentration. When cells are grown in batch culture, with ammonium as nitrogen source, most organisms possess glutamate dehydrogenase for the assimilation of the ammonium. The same enzyme is used when the micro-organism is cultivated with a growth-limiting carbon and energy source in a chemostat. These situations have in common that ammonia concentrations are relatively high. If, however, growth is restricted by using ammonium as growth-limiting substrate (low ammonia concentrations), the enzyme glutamine synthetase is induced (Figure 5.10).

Glutamate dehydrogenase is present as a constitutive enzyme. Glutamine synthetase has a considerably lower k_M (about 5 to 10 fold lower) and therefore higher affinity for NH_3 then glutamate dehydrogenase. Although the assimilation of NH_3 via glutamine synthetase requires 1 ATP it will be clear that, with ample energy available, this poses no problem for the cell. The enzyme glutamine synthetase is covalently modified at high ammonia concentrations which leads to inactivation of the enzyme.

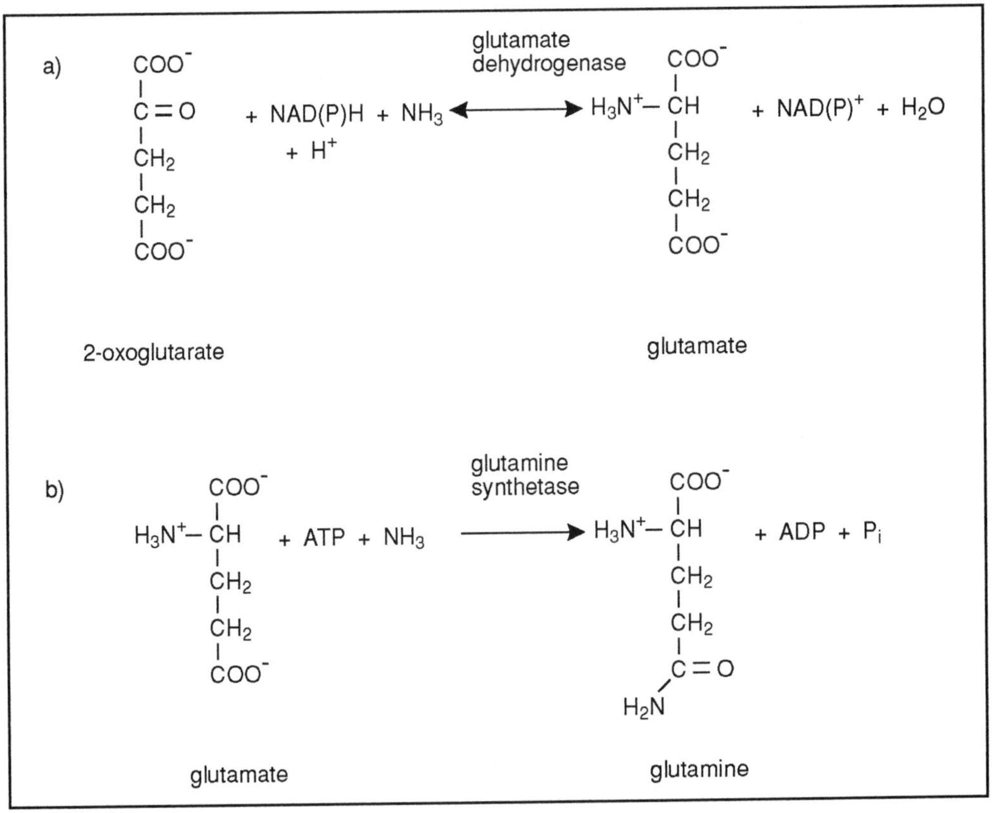

Figure 5.10 Ammonium as N source a) and as limiting nutrient b).

A very interesting phenomenon has been found in the lactic acid bacterium *Lactobacillus casei*, which, at high growth rates, ferments glucose to lactic acid. When grown in a chemostat with limited glucose supply and high dilution rate the bacterium produces mainly lactic acid however if the dilution rate is reduced the density of the culture increases and acetate appears in the culture liquid. This effect is shown in Table 5.3.

Dilution rate (h^{-1})	Products formed (mole/mole glucose)				C-recovery (%)	ATP formation (mole/mole glucose)
	Lactate	Acetate	Ethanol	Formate		
0.125	0	0.99	0.99	1.70	94	2.99
0.140	0	1.06	-	1.56	-	3.06
0.159	0.20	0.88	0.87	1.86	99	2.88
0.168	0.05	1.05	0.94	1.76	98	3.05
0.244	0.02	0.98	0.92	1.63	92	2.98
0.290	0.22	0.90	-	1.58	-	2.90
0.390	0.7	0.68	0.60	1.14	97	2.68
0.400	0.7	0.68	-	0.88	-	2.68
0.500	1.5	0.20	0.13	0.50	94	2.20
0.503	1.5	0.13	-	-	-	2.13
0.600	-	0.24	-	0.28	-	2.24

Table 5.3 Product formation, C-recovery and ATP formation for *Lactobacillus casei* grown at different dilution rates in a chemostat, with glucose as limiting factor.

Π   You might find it helpful to plot a graph of these data using [product] against dilution rate. It will make sure that you have understood the data.

This phenomenon can be explained by the fact that at low dilution rate *Lactobacillus casei* does not reduce pyruvate to lactic acid but, instead, transfers pyruvate to acetyl CoA. From acetyl-CoA, one ATP can be formed, along with acetate, formate and ethanol. So, depending on the growth rate, the following reactions take place:

At high μ

$$glucose \rightarrow 2 \text{ pyruvate} + 2 \text{ ATP} + 2 \text{ NADH}$$

$$2 \text{ pyruvate} + 2 \text{ NADH} \rightarrow 2 \text{ lactate}$$

$$\text{Overall reaction: glucose} \rightarrow 2 \text{ lactate} + 2 \text{ ATP}$$

At low μ

$$glucose \rightarrow 2 \text{ pyruvate} + 2 \text{ ATP} + 2 \text{ NADH}$$

$$2 \text{ pyruvate} + 2 \text{ NADH} \rightarrow 2 \text{ formate} + 1 \text{ ethanol} + 1 \text{ acetate} + 1 \text{ ATP}$$

$$\text{Overall reaction: glucose} \rightarrow 2 \text{ formate} + 1 \text{ ethanol} + 1 \text{ acetate} + 3 \text{ ATP}$$

We can see from Table 5.3 that at high μ the micro-organism is unable to produce energy as efficient as at low μ. The reason for this is not known yet but probably the rate of the reaction by which 3ATP is produced is not high enough to enable a high μ. It has been found that lactate dehydrogenase is activated by fructose-1,6-bisphosphate. At low growth rate the concentration of this glycolytic intermediate is low and lactate dehydrogenase does not function. Because of this pyruvate can be converted to acetate, formate and ethanol.

ecological studies

The chemostat is well suited to serve as a model system for studies on interactions between (aquatic) microbial populations. In most aquatic habitats nutrient concentrations are limiting and hence populations must compete for the available nutrients. Such studies have demonstrated that the success of a microbial population depends on substrate concentration and affinity for growth limiting substrate. Figure 5.11 shows this relationship for two different micro-organisms. Generalising one can say that some organisms are better able to grow at lower substrate concentration (K_s value relatively low) and hence are more successful in particular niches. So, selection based on competition for growth limiting substrate is a possible explanation for diversity within a microbial community. These kind of studies (using low substrate concentrations) would be impossible in batch cultures.

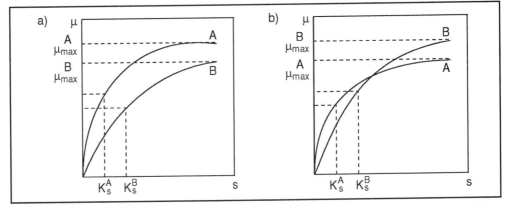

Figure 5.11 Growth rate (μ) at different substrate concentrations (s) for two different micro-organisms (A and B). As shown, the μ-s relationships for the organisms are in a) $K_s,^A < K_s,^B$ and $\mu_{max}^A > \mu_{max}^B$ and in b) $K_s,^A < K_s,^B$ and $\mu_{max}^A < \mu_{max}^B$.

If the conditions in a chemostat are kept constant, and the microbial niches overlap completely (compete for the same growth limiting substrate) eventually one population will survive whereas the others will be washed out. If more populations are to be kept in the same chemostat the conditions should be varied periodically so that alternate populations are in a favourable position. A multiple population can also exist in a chemostat when the populations are not limited by the same substrate. By varying the conditions in the chemostat it is thus possible to simulate natural circumstances. By using results from chemostat experiments one can predict to a certain extent which changes in population composition might occur in response to change in the natural habitat.

SAQ 5.5	1) Which of the three micro-organisms A, B, and C will become dominant in a competition experiment in a chemostat where the following steady-state concentrations ($\bar{s}$) of the rate limiting nutrient exist: 0.33×10^{-4}, 1.4×10^{-3} or 0.45×10^{-2} mol l^{-1}. The micro-organisms have the following μ_{max} and K_s: A) $\mu_{max} = 0.15$ h^{-1}; $K_s = 2 \times 10^{-4}$ mol l^{-1}. B) $\mu_{max} = 0.45$ h^{-1}; $K_s = 0.9 \times 10^{-3}$ mol l^{-1}. C) $\mu_{max} = 0.45$ h^{-1}; $K_s = 1.03 \times 10^{-3}$ mol l^{-1}. (Hint: consider equation 5.3).
	2) Are there any conditions in which micro-organisms C would dominate in the mixed culture? Explain your answer.

chemostat
enrichment
cultures

In Chapter 2 we considered enrichment cultures for the isolation of a particular micro-organism. A chemostat can also be used for this purpose. The chemostat is inoculated with a suitable inoculum and the culture grown at a specific dilution rate. Obviously the micro-organism which grows best at this dilution rate is favoured and the chemostat is thus useful for selecting rapidly growing species. An interesting example is the isolation of a *Pseudomonas* strain which can degrade 2,4,5-T, a herbicide extremely resistant to degradation by micro-organisms. Competition for 2,4,5-T supplied at very low level selected *P. cepacia* which can use the herbicide as sole energy and carbon source.

A disadvantage of enrichment in a chemostat is that slow-growing micro-organisms may be missed altogether, whereas such micro-organisms may play a very important role in nature.

microbial
metabolites

In principle it should be possible to adapt conditions in a chemostat in such a way that a maximum amount of a desired metabolite is produced, and yet this application has found only limited use. The main problem is that of genetic instability of the producing organism. Commercial micro-organisms used for product formation are usually highly mutated in order to produce a high yield of the desired product. This usually means that they are in physiological terms, very inefficient. A revertant strain developing in the culture is often better adapted to the conditions in the culture (ie more efficient) and eventually dominates the more productive mutant. This phenomenon has been called contamination from within and is the main reason why continuous culture is not used for the industrial production of microbial metabolites.

∏ A chemostat has been running in steady-state when suddenly the biomass concentration starts to increase again. Give possible explanations for this.

The increase in biomass can be explained by the following processes:

- contamination has occurred by another micro-organism with a higher yield in the chosen conditions;

- a mutant with a higher yield has developed within the culture;

- a different enzyme system has developed (not by mutation but otherwise for example by induction) that utilises the substrate more efficiently.

beer brewing

Although the brewing industry in the UK has used continuous culture for the production of beer on a commercial scale, this method has now been almost totally abandoned and replaced by improved batch systems. An advantage of continuous culture for brewing is that the fermentation time is only 4-8 hours compared to a week for the traditional batch system. However, with the introduction of the specially designed cylindroconical batch vessel the fermentation time for batch culture has been reduced to about 48 hours. Although this is still longer than the time needed for fermentation in continuous culture there are some definite disadvantages to the latter system which makes the industry prefer batch brewing. The most important disadvantages are:

- the long start-up period to reach the steady-state;

- the difficulty in matching the flavour of the continuous product with that of the traditional batch product.

waste water purification
Chemostat technology is applied on a large scale in the purification of (industrial) waste water, and is especially important when the chemical composition of the waste is fairly constant and comprises one or more (easily) degradable organic compounds. In some cases it is possible to make use of the conversion of these organic substrates to biomass for the production of so-called single cell protein, which can be used for animal feed. In other cases chemostat technology proves to be useful for the removal of (degradable) pollutants, thus preventing pollution of surface water by industrial waste water drainage.

5.8 Elaborations of the chemostat

biomass feedback

chemostats in series
Elaborations of the chemostat offer advantages in certain circumstances over the conventional chemostat. In this section we will consider two types of elaboration: biomass feedback and chemostats in series. These elaborations have extended the range of applications for continuous culture.

5.8.1 Chemostats with biomass feedback

biomass feedback systems
Biomass feedback refers to increasing the concentration of the biomass in the culture vessel. This is achieved by fitting some device, either internally or externally, to the chemostat. Figure 5.12 shows a variety of ways by which biomass concentration can be achieved. Use the following description to help you follow the details of this figure.

In system a) a cell-free or dilute biomass stream is removed through a filter and there is also an outlet for the concentrated biomass suspension. In system b) the culture is divided into a growth zone and a sedimenting zone by a baffle plate (metal disc with holes). Growth occurs in an agitated and homogenous zone below the plate. Above the plate there is a non-agitated sedimentation zone, virtually without growth, in which the biomass can sediment and return to the agitated zone. In this system a cell-free or dilute stream of biomass is withdrawn from the top of the fermentor and the concentrated biomass suspension leaves from an outlet in the growth zone. In system c) the 'monostream' system - it is assumed that the biomass is concentrated by sedimentation in the outlet pipe removing culture from the top of the fermentor. You should note that in this system there is no outlet for the concentrated biomass.

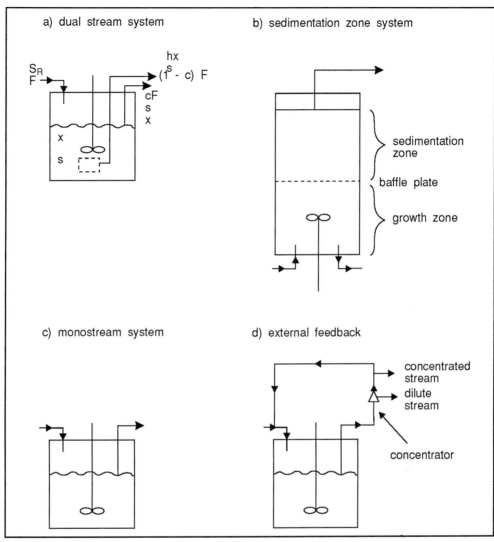

Figure 5.12 Various systems for feedback of biomass in a chemostat. Symbols F = flow rate, c = fraction of the outflow as a concentrated stream, (1 - c) = fraction of the outflow as a dilute stream, x = biomass concentration in the vessel, s = substrate concentration in the vessel, S_R = substrate concentration in the feed, hx = concentration of biomass in the dilute stream. These terms are explained more fully in the text.

∏ Can you think of factors that would influence the extent of biomass feedback in the monostream system?

Since biomass feedback is determined by sedimentation in the outlet pipe, the width of the outlet pipe and the flow rate will influence the extent of biomass feedback.

In system d) the biomass is concentrated outside the fermentor (external feedback system) and the culture is separated into two streams of biomass, one dilute and one concentrated; a part of the concentrated stream is fed back to the fermentor.

The mathematical relationships describing the performance of the systems shown in Figure 5.12 are different for each one. To illustrate the logic of how these relationships can be derived, we will consider only one biomass feedback device, the dual stream internal filtration system [a] in Figure 5.12]. In this system the fraction of the outflow which is not filtered is c. The outflow rate of the filtered biomass stream is therefore (1-c) F. The concentration of biomass in the filtered stream is h x; where h is related to the efficiency of biomass filtration ie when h = 1 there is no filtration and therefore no biomass feedback, when h = 0 the filter removed all biomass from the stream.

We already know that in a simple chemostat:

$$\frac{dx}{dt} = \text{growth - output} = \mu x - Dx$$

(E -5.2)

So with feedback

$$\frac{dx}{dt} = \text{growth - output in concentrated stream - output in dilute stream}$$

$$\frac{dx}{dt} = \mu x - cDx - (1\text{-}c)\, Dhx$$

In steady-state, $\frac{dx}{dt} = 0$, and after rearranging we have:

$\mu = [c\,(1\text{-}h) + h]\, D$

or $\mu = A\, D$ where $A = [c\,(1\text{-}h) + h]$

∏ We know that in a simple chemostat D = F/V (where V = culture volume). Is this true for the dual stream biomass feedback system (Figure 5.12a)? Explain your answer.

We can from Figure 5.12a that:

Flow in = output not filtered + output filtered

F = cF + (1-c) F

ie this becomes F = F, so, it is true to say that D = F for this system of biomass feedback.

∏ If the fraction of outflow which is not filtered is 0.8 determine the specific growth rate (μ) when D = 0.1h^{-1} and the efficiency of the biomass filter is 50% (Attempt this on a piece of paper before reading on).

c = 0.8, D = 0.1, h = 0.5.

$\mu = [c\,(1\text{-}h) + h]\, D$
$\mu = [0.8\,(1\text{-}0.5) + 0.5]\, 0.1$
$\mu = 0.09\ h^{-1}$

We can see that if c = 1 or h = 0 then there is no feedback and μ = D. So the limits of A are c to h and it follows that with feedback μ <D.

The growth-limiting substrate balance is the same as that for a simple chemostat (Equation 5.7). In the steady-state when $\mu = A\,D$ we have:

$$\bar{x} = (S_R - \bar{s})\,Y/A$$

We can see from the equation that biomass feedback increases the biomass concentration by the factor $\frac{1}{A}$; this is called the concentration factor.

The biomass output rate for unit volume of culture (productivity) in steady-state is again given by $D\,\bar{x}$.

The influence of biomass feedback on $\bar{x}$, D_c and output rate is shown in Figure 5.13.

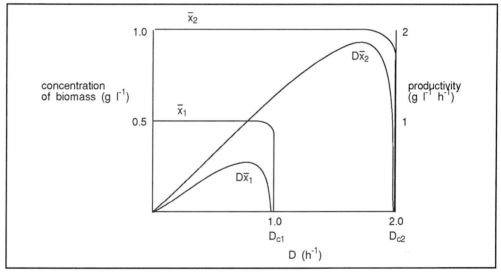

Figure 5.13 Comparison of biomass concentration and output rates in steady states of chemostat cultures with and without feedback. $\bar{x}_1$ = biomass concentration in chemostat culture without feedback; $\bar{x}_2$ biomass concentration in chemostat with feedback; $D\,\bar{x}_1$ = biomass output rate per unit volume without feedback; $D\,\bar{x}_2$ = biomass output rate per unit volume with feedback. D_{c1} = critical dilution rate without feedback. D_{c2} = critical dilution rate with feedback. Note that $\bar{x}_2/\bar{x}_1$ = concentration factor = $1/A$.

∏ We can see from Figure 5.14 that D_c with biomass feedback is increased far beyond that possible with a simple chemostat. Is μ_{max} also increased with biomass feedback? Give a reason for your response.

The correct answer is no, because μ_{max} for a given medium and given set of incubation conditions (such as pH and temperature) is set by the genetic constitution of the cell. Remember that with biomass feedback $\mu = A\,D$, so $\mu_{max} \approx A\,D_c$. ie D_c is extended beyond μ_{max}.

The main advantage of biomass feedback is that the maximum output rate of biomass (and products) in a chemostat with a given medium can be increased. For a fixed dilution rate, the fold increase will be the same as the concentration factor ($\frac{1}{A}$). This is useful when the growth-limiting substrate is unavoidably dilute eg in beer brewing and sewage purification, where a concentration factor of 100 has been achieved. Remember not only is $\bar{x}$ increased but it also means that we can increase D without getting wash out. Thus we can treat a large amount of sewage for example in a fairly small chemostat. Biomass feedback is also advantageous where the substrate has a low solubility or when the concentration of growth-limiting substrate has to be limited because of the formation of an inhibitory products. In addition, biomass feedback protects against 'shock loading' with an inhibitory substrate because the critical dilution rate is raised.

SAQ 5.6

From Figure 5.13 determine:

1) μ_{max} for the organism (assume the value of K_s is very small);

2) the concentration factor during biomass feedback;

3) μ at a steady state dilution rate of $1.5\,h^{-1}$ when biomass feedback is operating;

4) the fraction of outflow which is not filtered, assuming a biomass filter efficiency of 80%;

5) maximal productivity attainable with biomass feedback.

5.8.2 Chemostats in series

The joining together of two or more chemostats in series (Figure 5.14) leads to a multi-stage process which may have different conditions in each of the stages.

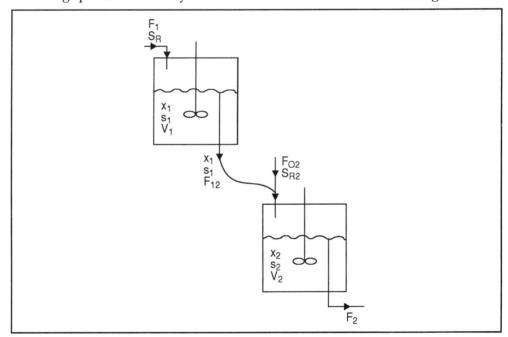

Figure 5.14 Two chemostats in series (multi-stream type). Note that if the system is in a steady-state, we can replace the biomass concentration x_1 and x_2 by their steady-state notations $\bar{x}_1$ and $\bar{x}_2$. Similar for substrate concentrations we could use $\bar{s}_1$ and $\bar{s}_2$.

single-stream
type

multi-stream
type

Chemostats in series may be of a single-stream type or a multi-stream type. In the single stream type medium is fed into the first stage of the series only. Here, for most steady-state dilution rates, the biomass concentration in the second stage $(\bar{x}_2)$ is the same as that for the first $(\bar{x}_1)$. However, the specific growth in the second stage (μ_2) is close to zero.

∏ Can you provide as explanation why $\mu_2 \approx 0$ for most dilution rates in a single-stream multi-stage system? What conditions would increase μ_2?

for
single-stream
type $\mu_2 \approx 0$

For most steady-state dilution rates the limiting substrate concentration in the first stage $(\bar{s})$, is maintained at a very low level (see Figure 5.4), so there is virtually no limiting substrate available for growth in the second stage, hence $\mu_2 \approx 0$. However, Figure 5.4 also shows that at dilution rates approaching μ_{max} there is an increase in $\bar{s}1$. Thus, more limiting substrate is available for second stage growth and μ_2 increases.

multi-stream
type

In the multi-stream type (Figure 5.14) fresh medium is fed to both first and second stages. In this system the overall dilution rate in the second stage (D_2) is given by:

$$D_2 = \frac{F_{o2}}{V_2} + \frac{F_{12}}{V_2}$$

$$D_2 = D_{o2} + D_{12}$$

where D_{o2} and D_{12} are partial dilution rates ie D_{o2} = dilution of second stage by flow (F_{o2}) of fresh medium; D_{12} = dilution of second stage by flow (F_{12}) from first stage.

The biomass balance in the second stage is given by:

$$\frac{dx_2}{dt} = \text{growth + input rate - output rate}$$

∏ Use appropriate symbols from those provided below to express: 1) second stage growth rate, 2) input rate and 3) output rate.

Symbols: D_c, μ_2, x_2, x_1, D_{12}, D_2, s_2, s_1, S_R

1) Growth $= \mu_2 x_2$

2) Input rate $= D_{12} x_1$

3) Output rate $= D_2 x_2$

So, $\dfrac{dx_2}{dt} = \mu_2 x_2 + D_{12} x_1, - D_2 x_2$

In steady-state, when $\dfrac{dx_2}{dt} = 0$, and after rearranging we have:

$$\bar{x}_2 = D_{12} \bar{x} / (D_2 - \mu_2)$$

The expression for $\bar{x}_2$ shows that as long as $\bar{x}_1$ is finite (greater than zero), $\bar{x}_2$ is finite, no matter how great the value of D_2. It follows that there is no critical dilution rate (D_c) for the second stage.

for multi-stream type $\mu_2 \approx \mu_{max}$

When the rate of addition of fresh medium to the second stage (D_{o2}) is fixed the specific growth rate in the second stage (μ_2) is maintained nearly at the maximum at all overall dilution rates (D_2; this is altered by adjusting D_{12}). Another feature is that over most of the range of dilution rate the biomass concentration is constant, and therefore stable, despite the high growth rate.

We will now consider some of the advantages that chemostats in series have over simple (single-stage) chemostats.

advantages of multi-stage systems

With single-stream systems for biomass production in which growth is limited solely by the supply of a single growth-limiting substrate there is virtually no advantage in using a series of chemostats. The reason for this is that over most of the possible ranges of dilution rates the growth-limiting substrate would be practically exhausted in the first stage and stationary phase conditions would occur in the second stage.

With complex media (eg with more than one source of carbon) a single-stream multi-stage process may be necessary to achieve utilisation of all substrate. For example, this would occur if utilisation of one carbon source represses the expression of genes for utilisation of another. In the single-stream multi-stage process, utilisation of one substrate in the first stage allows utilisation of the other in the second stage. In general, the system provides a series of different environments.

The multi-stream multi-stage system is a valuable means for obtaining steady-state growth when, in a simple chemostat, the steady-state is unstable eg when the growth-limiting substrate is also a growth inhibitor. This system can also be used to achieve stable conditions with maximum growth rate, an achievement which is impossible in a simple chemostat.

SAQ 5.7

Match appropriate continuous culture systems (a to e) with each of the features listed 1 to 3.

Feature

1) Steady states stable at high growth rates (μ close to μ_{max}).

2) Steady states stable at μ almost $= 0$.

3) Steady states can be achieved when D approaches μ_{max}.

Continuous culture system

a) Simple chemostat.

b) Turbidostat.

c) Dual-stream internal biomass feedback system.

d) Single-stream multi-stage system.

e) Dual-stream multi-stage system.

5.9 Some practical aspects of chemostat technology

In this section some practical aspects of cultivating micro-organisms in continuous culture will be briefly discussed.

culture stability

When we considered the kinetics of continuous culture (Section 5.4) we saw that the system is highly selective and favours propagation of the best-adapted organism present. In this case best-adapted refers to the micro-organism's affinity for the limiting substrate at the operating dilution rate. If, during the cultivation, a spontaneous mutation produces a faster-growing mutant then it will eventually outnumber the original micro-organism. You will recall (Section 5.7.2) that genetic instability limits the usefulness of chemostat culture for industrial production.

maintenance of aseptic conditions

When using a continuous culture which needs to function at steady-state conditions for longer periods of time it is very important to maintain aseptic conditions (that is avoid introduction of other micro-organisms). In chemostats this aspect needs much more attention than in batch cultures. For that reason a chemostat should be equipped with technical facilities enabling aseptic inoculating, sample-taking, medium supply, aeration, etc.

maintenance of steady-state conditions

To monitor establishment of steady-state conditions and to ensure they persist, certain parameters are determined either continuously or at regular intervals. These parameters are usually biomass and substrate concentration, pH, CO_2 and O_2 concentration (either in the culture liquid or in the exhaust gases), redox potential, concentration of products formed, etc. We will be discussing some of the techniques for measuring these in Chapter 7. Of course, probes that continuously sense parameters inside the culture vessel must be sterilisable (preferably *in situ*, by steam) and stable for long periods of time. Probes meeting these requirements have been developed for pH, redox potential, and dissolved CO_2 and O_2. There are several techniques for analysing the exhaust gases to determine their content of CO_2, O_2, NH_3, N_2, ethanol and other volatile compounds. If no probe is available for a certain compound which can be applied *in situ*, determination of that compound will have to be carried out in samples of the culture liquid itself.

Sometimes situations arise in chemostats which do not conform with theory. For example: micro-organisms of low μ persist in mixed cultures with micro-organisms of much higher μ, or, cultures are able to grow in a chemostat at dilution rates which are much higher than μ_{max}. Such discrepancies are often caused by so-called wall growth;

wall growth

development of a biofilm on the surfaces in the culture vessel. Cells adhering to surfaces grow in a micro-environment which can be different from the environment in the culture vessel itself. These adhering cells are not washed out and provide continuous inoculation of the medium, even at dilution rates higher than μ_{max}.

An added problem of biofilm-development is the fact that not only the vessel's walls are covered but also the probes. Consequently, these no longer function properly and control of steady-state conditions is lost.

mixing and aeration

A final aspect of continuous culture that must be considered is the need for proper mixing so that the culture liquid becomes sufficiently homogenous. In addition, when growing aerobic micro-organisms a sufficient oxygen supply must be maintained. This can be rather cumbersome because of the low solubility of O_2. Indeed, optimal O_2 concentration is not always feasible, especially at high cell density or in the case of filamentous micro-organisms (eg fungi), causing the culture liquid to become extremely viscous which hampers proper mixing. Actually, these problems are not only typical of growth in continuous culture, they are also important considerations in batch and fed-batch fermentor technology (see Chapter 7).

SAQ 5.8

For each of the following applications, select one or more appropriate culture system from the list provided.

Applications

1) Antibiotic production which occurs after growth has ceased.

2) A study of the influence of growth rate on the expression of a bacterial enzyme.

3) A process which relies on the operation of a metabolic pathway subject to catabolite repression from the limiting nutrient.

4) Biomass production using an unavoidably dilute substrate.

5) Selection of a fastest growing strain from a mixed population growing on the same limiting nutrient.

Culture systems

Simple chemostat.
Turbidostat.
Dual-stream multi-stage.
Single-stream multi-stage.
Chemostat with biomass feedback.
Repeated fed-batch.

Summary and objectives

In this chapter we have been seen that a chemostat provides a means of independently controlling growth rate and is a unique instrument for studying the physiology of micro-organisms. Growth is controlled by the rate of provision of a limiting nutrient and mathematical relationships can be used to predict growth and to determine growth parameters, such as μ_{max}, K_s, Y and m.

At dilution rates close to μ_{max} a turbidostat mode of operation can be used to stabilise the steady state. Here the biomass concentration is kept constant. We have also discussed fed-batch cultures and elaboration of chemostats.

Now that you have completed this chapter you should be able to:

- list the differences between growth in a batch and in continuous culture;

- understand and apply the mathematical description of growth in a chemostat;

- state the relationship between the specific growth rate and substrate concentration (Monod equation);

- write the equations in which the steady-state substrate and biomass concentrations can be calculated;

- describe the way in which the growth constants (K_s, μ_{max} and Y) can be determined experimentally;

- describe the influences of maintenance energy on the yield;

- apply the terms steady-state, dilution rate, growth limiting substrate, yield, critical dilution rate, Monod constant, productivity, maintenance coefficient and turbidostat;

- describe the principles of fed-batch, biomass feedback and multi-stage cultivation;

- give examples of the applicability of the different types of microbial cultivation techniques (batch, fed-batch, chemostat, turbidostat, biomass feedback, multi-stage);

- describe the main practical problems encountered in chemostat operation.

Environmental factors influencing growth

Environmental factors influencing growth

6.1 Introduction

In the previous chapters we examined growth of micro-organisms under optimal, constant conditions and by doing so we were able to define certain growth characteristics of micro-organisms. In nature however, growing conditions are usually far from stable and optimal, but vary a great deal, favouring some micro-organisms and at the same time inhibiting others. It follows that the composition of populations in natural habitats will, to a large extent, be determined by the physical and chemical characteristics of that environment. Micro-organisms can be found almost everywhere on earth, and even in extreme conditions growth has been found possible. Examples of these so-called extremophiles are *Sulpholobus* which grows in sulphur-rich hot acid springs at temperatures of up to 90°C and at pH values of 1-5, and *Halobacterium*, an extreme halophile which inhabits highly saline environments such as solar salt-evaporation ponds and heavily salted foods.

There are many environmental factors influencing growth, not only of micro-organisms but of all organisms. In this chapter we will consider some non-biological factors: temperature, availability of water, pH, oxygen, hydrostatic pressure and radiation. Knowing the (in)tolerance for these factors enables us either to combat undesirable micro-organisms or stimulate growth of useful ones. Some of this chapter should be a revisionary exercise as it develops, in detail, some of the discussions of Chapter 2 of this book.

6.2 Temperature

The effect of temperature on the growth rate of micro-organisms partly reflects the effect of temperature on the rate of (bio)chemical reactions.

The reaction rate of chemical reactions increases by a factor 1.8 for every 10°C rise in temperature. A similar effect can be seen in the influence of temperature on the growth rate (Figure 6.1).

At B (irreversible) denaturation of enzymes (proteins) sets in and the increase in reaction rate is counteracted by the negative effect of higher temperatures on the enzyme activity until finally no reactions can occur (C). It is believed that denaturation of proteins is not the only effect of high temperatures. It is very likely that membranes (which contain proteins and lipids) are affected as well.

cardinal
temperatures

There are three cardinal temperatures in the temperature/growth rate curve of Figure 6.1:

- a minimum temperature, below which there is no growth (A);

- an optimum temperature, at which growth is most rapid (B);

- a maximum temperature, above which there is no growth (C).

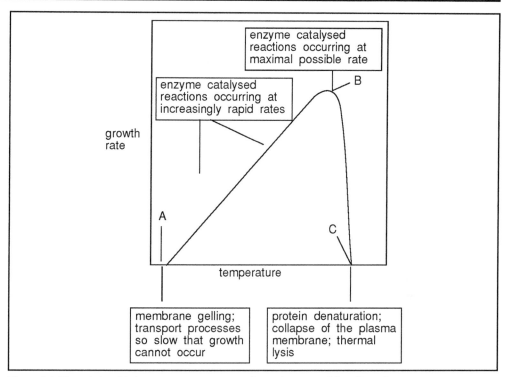

Figure 6.1 The relationship between temperature and growth rate, and the effect of temperature on different cellular components.

Π Can you explain why the optimum temperature is always nearer the maximum temperature than the minimum temperature for growth?

At temperatures above the optimum the influence of temperature on the rate of denaturation processes is stronger than the influence on the rate of biochemical reactions.

The temperature tolerance range of the microbial world reaches from temperatures well below freezing point to above 100°C. Broadly speaking micro-organisms can be divided into four groups: those with a low-temperature optimum, the *psychrophiles*, those with a mid-range optimum, the *mesophiles*, those with an optimum growth rate at high-temperatures, the *thermophiles* and those micro-organisms which grow best at extremely high temperatures and are incapable of any growth at low or mid-range temperatures, *extreme thermophiles*. Examples of these four groups and their temperature ranges are presented in Figure 6.2.

For each of the four groups of temperature tolerance, we will now consider the types of organisms and their natural habitats.

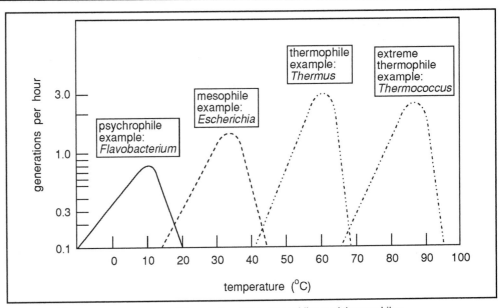

Figure 6.2 The temperature ranges of the psychrophiles, mesophiles and thermophiles.

6.2.1 Psychrophiles

There are large areas in the world with low temperatures almost all the year round: the Arctic and Antarctic (most of the time below 0°C), the oceans (average temperature 5°C and deeper down, 1 to 2°C), and the higher snow-covered mountains. In these areas micro-organisms can still be found as long as there is liquid water available. They belong to the group of *psychrophiles* and have their optimum growth rate at 15°C or lower, and their minimum at 0°C or lower.

psychrophiles

Permanent snowfields and glaciers often contain large quantities of psychrophilic algae, colouring the snow or ice surface distinctly red or green. The commonest of these algae is *Chlamydomonas nivalis*, producing bright red spores. These can be seen particularly in the second half of summer, in sunny dry areas, when the top layer of the snow melts or evaporates, revealing both spores and the green-pigmented vegetative cells.

The geographical areas already mentioned are cold the whole year round but there are also areas with low temperatures for only part of the year. These are the vast continental areas where the summer temperatures rises into the thirties, followed by very low temperatures in winter. Micro-organisms tolerating these low temperatures are not considered to be true psychrophiles: they have their optimum at 25-30°C, with a maximum at 35°C but are still able to grow (although only slowly) at 0°C. They are known as *psychrotrophs* or *facultative psychrophiles*.

psychrotrophs

Psychrotrophs have a much wider distribution than psychrophiles. They are found in soils and water in areas with temperate climates where the high summer temperatures are a strongly selective force excluding the obligate psychrophiles. A great variety of bacteria, fungi, algae and protozoa have members belonging to the psychrotrophs.

Spoilage of meat, milk and other dairy products, vegetables and fruits kept in the refrigerator, can be due to growth of psychrotrophic micro-organisms. Microbial growth is impossible in deep-frozen foods.

6.2.2 Mesophiles

Escherichia coli is a mesophile

To this group belong micro-organisms that live on man and warm-blooded animals, and in terrestrial or aquatic habitats in temperate and tropical climates. The temperature range for *mesophiles* is between 20 and 40°C. *Escherichia coli* is a well-known example. Many mesophiles have a temperature optimum of 37°C, the temperature of the human body.

∏ Replace letter a) to e) in the statements shown below with appropriate numbers or terms.

Usually a micro-organism is called psychrophilic if the optimum temperature is a) °C or lower and the minimum temperature below b) °C. Micro-organisms with an optimum temperature between 25 and 30°C and still growing at 0°C belong to the group of c) or d). The optimum temperature for e) is higher, between 35 and 40°C.

a) 15°C, b) 0°C, c) psychrotrophs, d) faultative psychrotrophs, e) mesophiles.

If you are not sure your response is correct re-read Sections 6.2.1 and 6.2.2.

6.2.3 Thermophiles

thermophiles

Whereas habitats for both psychrophiles and psychrotrophs can be found in large areas all over the world, suitable habitats for thermophiles are restricted to well-defined areas. Temperatures of about 50 to 70°C, needed for optimal growth of these micro-organisms, are for instance found in soils which are continuously heated by sunlight, or on a smaller scale in compost heaps and silage. The temperature in areas with volcanic activity, in hot springs or in steam vents is usually much higher, up to 100°C for hot springs and 350°C and beyond for steam vents.

Hot springs can be found in many places of the world (Central America, United States, Iceland, Japan, Indonesia, New Zealand, Central Africa). The water is usually at boiling point (boiling temperature depending on altitude) and overflows the edges of the spring creating a temperature gradient along the surrounding surface. Different micro-organisms develop in the various temperature ranges along the gradient presenting the microbiologist with the possibility of determining the upper temperature limit for each type of organism (see Table 6.1).

From such studies certain conclusions can be drawn:

- in general prokaryotic micro-organisms are able to grow at higher temperatures than eukaryotes;

- non-photosynthetic organisms are able to grow at higher temperatures than photosynthetic organisms.

However, one must bear in mind that these conclusions do not apply for all micro-organisms of the three types mentioned: the possibility of growing at the higher temperatures is restricted to only a few extremely thermophilic species or genera.

Group	Approximate upper temperature (°C)
Eukaryotic micro-organisms	
Protozoa	56
Algae	55-60
Fungi	60-62
Prokaryotic micro-organisms	
Cyanobacteria (blue-green algae)	70-73
Photosynthetic bacteria	70-73
Chemolithotrophic bacteria	> 100
Heterotrophic bacteria	> 100

Table 6.1 Upper temperature limits for growth of micro-organisms.

∏ What general conclusion can be drawn from the studies on the temperature limit for micro-organisms (Table 6.1)?

Structurally simple organisms can grow at higher temperatures than more complex organisms.

extreme thermophiles

Many of the *extreme thermophiles* belong to the group of *Archaebacteria*. Together the extreme thermophilic, acidophilic, alkalophilic and halophilic (salt loving) micro-organisms are known as *extremophiles*. Some bacteria can be included in this group on the basis of more than one property (for instance those that are able to grow at pH 2.0 and 80°C).

It is assumed that the upper temperature limit at which microbial life is still possible is lower than 150°C. An interesting feature of thermophiles is their quite high growth rates. This is due to fact that their optimum temperatures are the same as the temperature of their environment.

Thermophiles are not restricted to naturally hot habitats. Micro-organisms resembling *Thermus aquaticus* (a common hot spring inhabitant) also thrive in man-made habitats such as domestic water heaters, water heated in industrial processes or water in cooling systems, for instance, in electric power plants.

SAQ 6.1

Doubling times for some species of bacteria growing at their optimum growth temperatures are shown below:

Bacterial species	Temperature, °C	Doubling time, hr
Vibrio vulnificus	30	0.16
Alteromonas haloplanktis	10	1.35
Escherichia coli	37	0.33
Neisseria sicca	37	1.24
Methanobacterium thermoautotrophicum	65	2.50
Nitrosomonas europaea	25	18.50
Bacillus cereus	40	0.43
Bacillus globisporus	60	0.22

1) Which species grew at the fastest rate?

2) Which, if any, species could be described as being thermophilic?

3) Which, if any, species are psychrophiles?

4) *E. coli* was also grown at 30°C and 40°C. What would happen to the doubling time at these temperatures?

5) 'All *Bacillus* species are thermophiles'. From the data, is this statement true or false?

6) Two species are suspected of being human pathogens. Which are the most likely candidates and why?

7) Which species is most likely to occur in a marine environment?

8) Given a temperature of 50°C, which species are most likely to grow?

9) At 27°C in a mixed population of *Vibrio* and *Nitrosomonas*, which species is likely to outnumber the other?

SAQ 6.2

An extremely thermophilic organism has been isolated from a submarine volcanic habitat. Its growth characteristics are shown below.

Temperature, °C	Doubling time, min
80	no growth
85	550
100	220
105	110
110	poor growth

1) What is the optimum growth temperature for this organism?

2) What is its maximum growth temperature?

3) What is its minimum growth temperature?

4) What temperature range is most likely to be encountered in this organism's habitat?

5) The organism could be re-isolated and grown from samples stored at 4°C over a period of at least 10 months. What can you deduce from this?

6.2.4 Factors determining temperature limits for growth

We have attributed the drop in growth rate at higher temperatures to denaturation of the proteins in the cell (Figure 6.1). Research on the factors determining the limits for growth has been carried out along two lines: 1) by comparing the properties of organisms with widely different growth limits, and 2) by studying the properties of temperature-sensitive mutants of which the temperature range has been decreased by only one single mutational change.

heat-stability of cell proteins

When comparing specific proteins of thermophiles with those of mesophiles many of the former appear to be much more heat-stable. The overall stability of soluble cell proteins can be determined by measuring the rate at which the protein in a bacterial extract becomes insoluble (precipitates) when denatured by heat at different temperatures. Such experiments show clearly that nearly all the proteins of a thermophilic bacterium remain in their native state at temperatures which denature most proteins of related mesophiles (Table 6.2). This leads to the conclusion that adaptation of thermophiles to their high-temperature environment is the result of mutational changes that affect the primary structure (amino acid sequence) of most, or maybe even all, of their proteins.

So the evolutionary changes that have lead to thermophilic organisms have caused an increase in the thermal stability of their proteins. This is in contrast with the observation that most mutations affecting the primary structure of a specific protein (enzyme) decrease its thermal stability although they may have little effect on the enzyme's catalytic properties. So if in natural circumstances there is no counterselective thermal pressure, the maximum growth temperature of any micro-organism should drop progressively as a result of random mutations affecting the primary structure of its proteins. Psychrophilic bacteria from the Antarctic have indeed been found to contain a large number of exceptionally heat-labile proteins.

Organism	Temperature class	Percent proteins denatured
Proteus vulgaris	Mesophile	55
Escherichia coli	Mesophile	55
Bacillus megaterium	Mesophile	58
Bacillus subtilis	Mesophile	57
Bacillus stearothermophilus	Thermophile	3
Bacillus sp. (Purdue CD)	Thermophile	0
Bacillus sp. (Texas 11330)	Thermophile	4
Bacillus sp. (Nebraska 1492)	Thermophile	0

Table 6.2 Stability of cytoplasmic proteins from thermophilic and mesophilic bacteria at 60°C. Data derived from Stanier, RY, Ingraham, J.L, Wheelis, ML and Painter, PR (1986) General Microbiolgy, MacMillan, Basingstoke UK.

lipid
composition

In nearly all organisms the lipid composition changes with the temperature. At lower temperatures the proportion of unsaturated fatty acids in the cell's lipids increases (Table 6.3).

	Temperature of growth	
Fatty acid	10°C	43°C
Saturated fatty acids		
Myristic (14.0)	3.9	7.7
Palmitic (16.0)	18.2	48.0
Unsaturated fatty acids		
Palmitoleic (16.1)	26.0	9.2
Oleic (18.1)	37.9	12.2

Table 6.3 Effect of temperature on the fatty acid composition of lipids in *Escherichia coli*. Data from Marr, AG and Ingraham, J.L, (1962) J. Bactenology 8.4, 1260.

This phenomenon is the cell's adaptation to lower temperatures since the melting point of lipids is directly related to the proportion of saturated fatty acids. Therefore, the degree of saturation of the fatty acids determines the fluidity of the lipids. Functioning of the membrane depends on the fluidity of the lipids so if growth is to take place at low temperatures the degree of unsaturation of fatty acids should be increased.

Indeed, psychrophiles contain a much larger quantity of unsaturated fatty acids in their membranes than mesophiles whereas in thermophiles the amount of saturated fatty acids is much higher. This larger content of unsaturated fatty acids provides the psychrophiles with a well-functioning active transport system (necessary to take up essential nutrients) even at low temperatures. Whereas the saturated fatty acids in the membranes of thermophiles improve the stability of the membranes since saturated fatty acids form much stronger hydrophobic interactions than unsaturated fatty acids.

Π Psychrophiles and thermophiles differ from mesophiles in their ability to grow at extreme temperatures. Try to describe which properties of the first two groups of micro-organisms are responsible for this.

To grow at extreme temperatures the cell should have a stable, functioning membrane (for intake of nutrients, proton gradient, etc). A functioning membrane is characterised by a certain fluidity. To achieve this psychrophiles have a higher percentage of unsaturated fatty acids than mesophiles, which leaves the membranes fluid even at low temperatures. This is as opposed to thermophiles which have membrane lipids rich in saturated fatty acids. Moreover, both types of organisms have an adapted amino acid composition of their proteins which makes them better suitable for life at extreme temperatures.

We saw in Table 6.1 that, in general, the maximum temperature for eukaryotes is lower than that for prokaryotes and extreme thermophiles are only found among bacteria. Presumably this is the consequence of the thermolability of the organelle membrane structure (nucleus, mitochondria, chloroplasts, etc).

effect of very low temperatures

The temperature range for growth of any one organism is about 30 to 40°C. The changes in the cell due to high temperatures are usually irreversible but at temperatures below the minimum they are often reversible. Bacteria can be kept for decades at liquid nitrogen (-196°C) temperatures provided they are frozen under the right conditions.

So, how can we protect cells from irreversible damage at freezing temperatures? Because of its high concentration of proteins and inorganic salts the cell content of micro-organisms has a high osmotic value. This lowers its freezing point to below 0°C and when cells are exposed to freezing temperatures their contents become supercooled. Depending on the speed at which the temperature is lowered and on the cells' permeability to water one of two things can happen to cells. If the temperature drops slowly and the cells are highly permeable to water they lose water and dehydrate. In the case of low permeability to water and rapid cooling ice crystals are formed inside the cells. Although dehydration is harmful enough to cause cell death the main damage of freezing is physical disruption of cellular structures, especially the cell membrane by the ice crystals formed. Cells can be protected from frost damage by the medium in which they are suspended. If glycerol or dimethylsulphoxyde, which can penetrate the cell, are added at about 0.5 moles l^{-1} the effects of dehydration are reduced. Glycerol can replace water in the watercore around proteins if water is removed from the cell. Certain macromolecules, such as serum albumin (protein), and dextran (a glucose polymer) also protect the cell. Since these compounds do not penetrate the cell, it is believed they act by somehow adhering to the cell membrane and thus protect it.

SAQ 6.3	Identify each of the following statements as true or false.

1) Thermophiles contain much more saturated fatty acids in their membrane than psychrophiles.

2) Some fungi are able to grow at 100°C.

3) Most mutations affecting primary structure of an enzyme increase its thermal stability.

4) In their natural environment thermophiles are not able to survive temperatures of 4°C for extended periods.

5) Growth rates at optimum temperatures for thermophiles are generally lower than for mesophiles.

6) Cells may be protected from irreversible damage at freezing temperatures by lowering the osmolarity of the cytoplasm.

7) Addition of dextran to the suspension medium reduces the effects of dehydration.

6.3 Water availability

6.3.1 Water activity

All organisms need water to live. In natural habitats the availability of water is a prime influence on growth. Although water may be present in an environment this does not necessarily mean that it is also available to the (micro)organisms. It may be absorbed by solid substances or surfaces so that it is unavailable. When water serves as solvent (which it frequently does) its solutes, such as salts and sugars, have a strong affinity for water which also makes it unavailable. So, micro-organisms often need to compete with their environment for the available water. If the availability is limited because of solutes, we speak of an osmotic effect. Adsorption at surfaces is called a matrix effect.

water activity
(a_w)

To define water availability we make use of the physical terms water activity or water potential. The former term is mostly used in food microbiology and it is expressed as the ratio of the vapour pressure of water in the air above the substance or solution and the vapour pressure of pure water at the same temperature. So the value of water activity (a_w) is always a figure between 0 and 1. Table 6.4 shows the water activity of several materials. Not mentioned are values for the water activity in agricultural soils which usually range between 0.9 and 1.0.

Water activity, a_w	Material	Some organisms growing at stated water activity
1.000	Pure water	*Caulobacter, Spirillum*
0.995	Human blood	*Streptococcus, Escherichia*
0.980	Seawater	*Pseudomonas, Vibrio*
0.950	Bread	Most Gram-positive rods
0.900	Maple syrup, ham	Gram-positive cocci
0.850	Salami	*Saccharomyces rouxii (yeast)*
0.800	Fruit cake, jams	*Saccharomyces bailii, Penicillium* (fungus)
0.750	Salt lake, salt fish	*Halobacterium, Halococcus*
0.700	Cereals, candy, dried fruit	Xerophilic fungi

Table 6.4 Water activity of several materials.

The water activity can be expressed by the following formula:

$$a_w = \frac{P_w}{P_w^o}$$

in which:

a_w = water activity

P_w = vapour pressure of water in the atmosphere over a solution of a compound in water at a certain temperature.

P_w^o = vapour pressure of pure water at the same temperature.

Raoult's law   For very diluted solutions Raoult's law may be applied:

$$\frac{P_w}{P_w^o} = \frac{n_1}{n_1 + n_2}$$

in which:

n_1 = number of moles water
n_2 = number of moles solute

So, the water activity can be calculated when the number of moles of the solute is known. However, this formula is only applicable with ideal solutions. The more polar the solute or the more dissociated (as will be the case for electrolytes) the greater the difference between its a_w value and that of an ideal solution. In such cases experimental water vapour pressures are used.

∏   Try to think of a few examples in your environment where the water activity is influenced by dissolved compounds or absorbing surfaces.

There are, of course, many different examples. Here are a few that you may, or may not, have thought of.

The influence on water activity by adding water soluble compounds can be found in a large number of foodstuffs with a concentration of salt or sugar. For example: marmalade and jams, syrup, treacle, molasses, marzipan, certain types of sausages or other salted meat and fish products. The water activity is usually not sufficiently low to prevent the food from going bad although many food spoilers are unable to grow in it.

It is known that because of their composition many natural materials are much more hygroscopic than their synthetic equivalents. This plays an important role in clothing and soft furnishings. Wool is especially hygroscopic which gives it a strong buffering effect on the relative humidity of the environment. This contributes to the sense of comfort usually experienced with this kind of material.

∏   Why does the water activity vary less with temperature than water vapour pressure does?

We know that water activity is defined as:

$$a_w = \frac{P_w}{P_w^o}$$

P_w and $P_w o$ vary in a more or less similar way with the temperature. As a consequence the quotient will be relatively independent of the temperature. This means that in a closed system the water activity will remain reasonably constant provided the temperature changes are not too great.

In soil microbiology the term water-potential is preferred to water activity. Water potential also expresses the availability of water, but is an energy term, defined as the difference in the Gibbs function (ΔG) between the system under study (for example, soil, food) and a pool of pure water at the same temperature. Water potential can be expressed in a number of different units, but the most widely used unit is the bar, which is equivalent to 0.986 atmospheres (atm). The values are always negative. Water activity and water potential are not exactly proportional, because water activity does not take into account the temperature of the system, but as a rough rule, a decrease of 0.01 a_w in the range 0.75 to 1.0 a_w is equivalent to a decrease of potential of about 15 bar.

∏   Clay absorbs water more strongly than that of sand. What can you deduce about the relative water potential of these materials?

The water potential of clay will be lower (more negative) than sand containing the same amount of water.

6.3.2 Osmotic tolerance

Osmotic tolerance describes the ability of an organism to grow in a medium of varying solute concentration. Micro-organisms achieve this by adapting the solute concentration of their cytoplasm so that this concentration is kept higher (hypertonic) than that of the surrounding medium. This draws water into the cell by diffusion and an osmotic pressure develops. Most micro-organisms can resist considerable osmotic pressure since they possess a rigid cell wall.

osmoregulation   The capacity of osmoregulation - maintenance of solute concentrations at levels above that of the environment, and optimal for metabolic processes - is limited for most micro-organisms in that they are unable to withstand environments of very low water activity (high solute concentration). If the internal osmotic pressure drops below that of the environment ie becomes hypotonic, water leaves the cell and the volume of the cytoplasm decreases. In Gram-positive bacteria the effect is withdrawal of the cell membrane from the cell wall which is known as *plasmolysis* and leads to damage of the cell membrane.

Some micro-organisms do not only have the capability to adapt but even prefer environments of low water activity. In general, yeasts and filamentous fungi grow better in such environments than bacteria (see Table 6.4). According to their preference, micro-organisms can be divided into three groups: halophiles, osmophiles and xerophiles.

halophiles, osmophiles and xerophiles   *Halophiles* prefer a high salt concentration, *osmophiles* a high solute (usually sugar) concentration, and *xerophiles* an environment of low moisture.

To be able to survive, these organisms rely on mechanisms that make sure the water activity in the cytoplasm is low (lower than that of the environment). There are several possible mechanisms:

- pumping ions from the environment into the cell;

- producing an organic solute inside the cell.

Whatever the solute used for this purpose, it should not be toxic. Suitable compounds are called compatible solutes.

compatible solutes   Examples of compatible solutes are:

- the amino acid proline (in *Staphylococcus*) and glutamate;

- inorganic ions. This is for instance the case for *Halobacteria* (for example *Halobacterium halobium*) which live in very saline habitats, such as salt lakes and salt evaporating ponds. They use potassium ions (K^+) which they pump into the cell. There are many other bacteria that accumulate K^+ in their cytoplasm rather than Na^+. In fact there is a direct correlation between the osmotic tolerance of several bacteria and their K^+ content. Experiments with *Escherichia coli* have shown that the intracellular K^+ concentration increases progressively with increasing osmolarity of the medium. The result is an increase in osmolarity and ionic strength of the cell.

- a third group of possible compatible solutes are the polyalcohols such as sorbitol and ribitol. By producing such compounds some yeasts can tolerate very high sucrose or salt concentrations. They pose problems by spoiling food which is rich in sucrose (syrups) or salt (for instance soy sauce).

At water activity values below 0.6 micro-organisms cannot grow, but they may still survive. This phenomenon is made use of when preserving micro-organisms by freeze-drying. Another example of survival by micro-organisms in a dry environment is when they are present in aerosols. Generally spore-forming micro-organisms and

Gram-positive cells keep their viability longer than Gram-negative cells. This is an important factor in the transfer of germs by air.

halophiles

Most natural environments of high osmolarity contain high concentrations of salts, particularly sodium chloride. According to their tolerance for this salt bacteria can be divided into four categories: *non-halophiles, marine organisms, moderate halophiles* and *extreme halophiles*. Examples are given in Table 6.5 and the effect of sodium ion concentration on growth of micro-organisms is shown in Figure 6.3.

Physiological class	Representative organisms	Approximate range of NaCl concentration tolerated for growth (% g/100 ml)
Nonhalophiles	*Aquaspirillum serpens*	0.0-1
	Escherichia coli	0.0-4
Marine forms	*Alteromonas haloplanktis*	0.2-5
	Pseudomonas marina	0.1-5
Moderate halophiles	*Paracoccus halodenitrificans*	2.3-20.5
	Vibrio costicolus	2.3-20.5
	Pediococcus halophilus	0.0-20
Extreme halophiles	*Halobacterium salinarium*	12-36 (saturated)
	Halococcus morrhuae	5-36 (saturated)

Table 6.5 Osmotic tolerance of certain bacteria. Data from Stanier, RY, Ingraham, J.L, Wheelis, ML and Painter, PR (1986) General Microbiology, MacMillan, Basingstoke UK.

Some halophiles, such as *Pediococcus halophilus*, can grow on a medium of very high salt concentration, but also on a medium without any added NaCl. Other bacteria, including marine bacteria and certain moderate halophiles, always require NaCl to grow. In general marine micro-organisms are inhibited at both high and low salt concentrations. They require sodium ions for two reasons: for the stability of the cell membrane and for the activity of many of their enzymes. Of the extreme halophiles *Halobacterium* is worthy of mention because of the construction of its cell wall. This consists only of glycoprotein (that means it does not have a layer of peptidoglycan) and yet it is rigid enough to give the cell its cylindrical shape in very concentrated salt solutions (25-35% w/v, ie g solute per 100 ml solvent). In lower salt concentrations (15% w/v) the cells become round and in still more diluted solutions the cell wall disintegrates into protein monomers and lysis occurs. This kind of experiment indicates that the structure of the cell wall in this bacterium is maintained by ionic bonds between the protein units, provided by high concentrations of Na^+ (which can not be replaced by other monovalent cations).

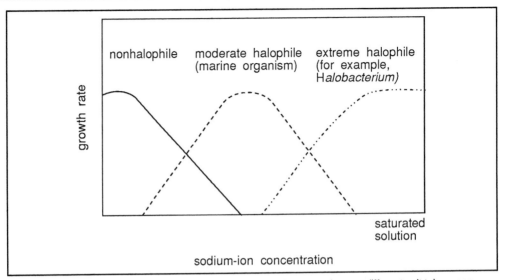

Figure 6.3 Effect of sodium ion concentration on growth of micro-organisms at different salt tolerances.

∏   Can you explain why yeasts and moulds are the main spoilage organisms of materials such as paint and wall paper on damp walls.

Due to the matrix effect these materials have a low water activity which favours the growth of yeasts and filamentous fungi. However, the composition of the material is also important: only those organisms that produce the necessary, often extracellular, enzymes can develop.

SAQ 6.4	Identify each of the following statements as true or false. If false give a reason for your response.

1) All halophiles are osmophiles but not all osmophiles are halophiles.

2) Extreme halophiles require greater than 20% NaCl to reproduce.

3) Moderate halophiles can tolerate NaCl concentrations up to about 20%.

4) Plasmolysis of cells may occur if the water activity of the cytoplasm is far higher than that of the environment.

5) Jams have a long shelf life because they are hypertonic compared to the cytoplasm of most cells.

6) All halophiles require some salt to grow.

7) Osmophiles are able to lower the water activity of their cytoplasm.

6.4 Influence of pH on growth

As you will know from chemistry pH is defined as the negative logarithm of the proton concentration of the solution. This implies that a difference in pH of one unit represents a tenfold difference in proton concentration. pH 7 means the substrate or solution is neutral. Solutions with pH < 7.0 are acidic and those with pH > 7.0 are alkaline.

All (micro) organisms have a certain pH range in which they can grow and within that range there usually is an optimum pH at which they grow best. Most natural environments have pH values between 5 and 9 and micro-organisms with their optimum pH in this range are most common. Only a few micro-organisms can live above pH > 9 (the so-called *alkalophiles*) or pH < 2 (the *acidophiles*). Examples of the latter group, (including some animals and plants) can be found in Table 6.6. To the alkalophiles belong certain cyanobacteria (*Spirulina*) that grow in the so-called soda lakes in Africa.

acidophiles
and
alkalophiles

Group	Approximate lower pH limit	Examples of species found at lower limit
Animals:		
Fish	4	Carp
Insects	2	Ephydrid flies
Plants:		
Vascular plants	2.5-3	*Eleocharis sellowana*
		Eleocharis acicularis
		Carex spp.
		Ericacean plants (heather, blueberries, cranberries, etc)
Mosses	3	*Sphagnum*
Eukaryotic micro-organisms:		
Algae	1-2	*Euglena mutabilis*
		Chlamydomonas acidophila
		Chlorella spp.
	0	*Cyanidium caldarium*
Fungi	0	*Acontium velatum*
Protozoa	2	Amoebae, heliozoans
Bacteria:	0.8	*Thiobacillus thiooxidans*
		Sulfolobus acidocaldarius
	2-3	*Bacillus, Streptomyces*
Cyanobacteria	4	*Mastigocladus, Synechococcus*

Table 6.6 Lower pH limits for different groups of organisms.

The table indicates that, on the whole, micro-organisms are more tolerant of low pH values, with the exception, however, of the cyanobacteria. These bacteria are hardly ever found in environments with pH < 4.

Acidophilic micro-organisms can be facultative or obligate acidophiles. Most fungi and yeasts are facultative acidophiles: they grow reasonably well at low pH, with an optimum pH at about 5, but they can also grow well at pH 7. Some algae are obligate acidophiles which can only grow well at low pH and not at all at pH around 7. Of the obligately acidophilic bacteria the genus *Thiobacillus* is a well-known example (its members cause the acidity of mine drainage). Other examples are *Sulpholobus* and *Thermoplasma*, which are both thermophilic as well. For such obligate acidophiles a neutral pH is usually toxic. This is caused by the effect on the cell membrane which actually dissolves. This implies that high concentrations of hydrogen ions are needed for maintaining the membrane structure.

∏ List the following groups in order according to their approximate lower pH limit, starting with the lowest.

Bacteria, Protozoa, Plants, Fungi, Algae, Fish.

Now check your order with the data given in Table 6.6.

maintenance of intracellular pH

Whatever the pH of the environment the pH of the cell's cytoplasm is usually kept near neutrality. This is the pH at which most cellular processes work best and the cellular compounds are most stable. Proteins, DNA and chlorophyll are sensitive to low pH, RNA and phospholipids are unstable at higher pH.

Maintenance of a neutral intracellular pH is also important for the cell's energy supply where this is generated by the proton motive force (PMF). According to Mitchell's chemi-osmotic theory the PMF originates from a difference in charge and a difference in pH across the membrane. So depending on the pH of the environment the contribution of pH to the PMF may vary quite considerably. It follows that the cell must have some mechanism to prevent H^+ from entering or pump H^+ out once it is in.

In laboratory cultures the pH of the medium can be strongly affected by metabolic products of the growing organism. Many fermenting bacteria produce organic acids (lactic acid, acetic acid, etc) and ammonia (released when amino acids are fermented). When culturing bacteria the pH of the medium is usually kept at the cell's optimum pH by buffering or by keeping the medium's pH stable in some other way. Depending on the pH required the most commonly used buffers are phosphate (pH 6-8), citrate (3-7), borate and glycine (7-9). Which buffer is chosen for which organism also depends on any possible toxic effect the buffer may have. It is, for instance, known that phosphate is toxic to some bacteria. If a buffer is not tolerated the medium can be kept at a certain pH by adding (either manually of by some automatic device) small quantities of either alkaline or acid, whichever is needed.

| SAQ 6.5 | Which of the following characteristics/processes would aid an acidophilic bacterium growing at pH 2.0? |

1) The cell membrane being relatively permeable to cations.

2) The cytoplasm having a relatively high buffering capacity.

3) Active transport of protons from the cell into the periplasm.

4) A small intake of protons by the cell membrane creating a positive charge inside the cell, which inhibits further inward transport.

6.5 Oxygen

role of oxygen in the cell

Together with a number of other elements oxygen plays a vital role in the living cell. The average microbial cell contains about 20% oxygen, which means that this element and carbon (about 50%), make up the bulk of the cell content. Nearly all of the cell's oxygen is derived from water, organic substrates or CO_2. Only when growing on certain substrates such as hydrocarbons, in which an oxygenase enzyme is involved, is molecular oxygen (O_2) incorporated in the cell. Some cell components, such as sterols, can only be formed when oxygen is present. During long-term anaerobic culturing of yeasts sterol often need to be added to the medium because the micro-organism is no longer able to synthesise this compound itself. Another important role of oxygen is that of final hydrogen and electron acceptor in aerobic respiration.

6.5.1 Division of micro-organisms according to oxygen tolerance

As we have seen for other factors, micro-organisms can be divided into several groups according to their need for or tolerance of oxygen. Table 6.7 shows the groups and their relation with oxygen.

Group	O_2 effect
Aerobes	
Obligate	Required
Facultative	Not required, but growth better with O_2
Microaerophilic	Required, but at levels lower than atmospheric
Anaerobes	
Aerotolerant	Not required, and growth no better when O_2 present
Obligate (strict) anaerobes	Harmful or lethal

Table 6.7 Division of micro-organisms according to their relation with oxygen.

aerobes and anaerobes

The aerobes are dependent on oxygen, while the facultative aerobes use oxygen when it is available but they can also do without. Anaerobes are unable to use oxygen. They can be divided into two groups: the obligate anaerobes, for which O_2 is toxic (lethal) and the aerotolerant anaerobes for which exposure to O_2 is not lethal. For aerobes, however, high O_2 concentrations may also be toxic. Many obligate aerobes cannot grow in O_2

concentrations higher than atmospheric (ie more than 20%) and some of them require O_2 concentrations well below that mark, namely 2-10 per cent v/v in order to grow. The latter kind of aerobes belongs to the so-called microaerophiles.

6.5.2 Toxic forms of oxygen

oxygen toxicity

Molecular oxygen (O_2), as it occurs in the atmosphere, differs from other diatomic elements because in its ground state the two outer electrons are in separate orbitals and have parallel spins. This means that the magnetic fields that are induced by the electrons as they turn around their axis have the same direction. For electrons with antiparallel spins the magnetic fields compensate each other. This makes oxygen in its ground state, the triplet form, paramagnetic. This is shown, in a simplified form, in Table 6.8. The consequence is, that oxygen is not very reactive with other atoms or molecules that do have an electron pair with antiparallel spin.

Form	Formula	Simplified Electronic Structure	Spin of Outer Electrons
Triple oxygen (normal atmospheric form)	$^3O_2$	Ȯ—Ȯ	⊕ ⊕
Singlet oxygen	$^1O_2$	Ȯ—Ȯ	◍ ○ or ⊕ ⊖
Superoxide free radical	O_2^-	Ö—Ȯ	◍ ⊕
Peroxide	O_2^{2-}	Ö—Ö	◍ ◍

Table 6.8 Some derivatives of oxygen and their electronic configuration.

singlet form is very reactive

By adding energy the outer electrons of oxygen can also obtain an antiparallel spin, in a separate or in the same orbital. This form, the singlet form, is very reactive and, because of that, toxic for the cell.

(Bio)chemically there are several ways in which singlet oxygen can be formed. The most common way is the reaction of triplet oxygen with visible light in which a light absorbing dye molecule acts as mediator. Other biochemical singlet oxygen generating reactions involve enzyme systems such as *lactoperoxidase* and *myeloperoxidase*. These enzyme systems are present in milk, saliva and in cells that eliminate invading micro-organisms by phagocytosis. When an invading micro-organism is ingested, the peroxidase system of the cell is activated and singlet oxygen formed, killing the invader. In peroxidase systems, chloride ion is converted to hypochlorite (OCl⁻), which reacts with hydrogen peroxide in the following way:

$$H_2O_2 + OCl^- \rightarrow {}^1O_2 + OH^- + HCl$$

Once present in a biological system singlet oxygen is very destructive, setting in motion oxidation reactions which can destroy vital cell components, such as phospholipids in the cell membrane.

Cells can be protected against singlet oxygen by the presence of light absorbing pigments called carotenoids which absorb the energy of the singlet form so that the triplet form is regenerated. This effect is well demonstrated by the relatively large number of pigmented micro-organisms that are found as airborne infections. Apparently non-pigmented cells have less chance of survival in the atmosphere. A good example is the potentially pathogenic bacterium *Staphylococcus aureus* which can survive bright sunlight quite well. Infection with this pathogen is often airborne and the bacterium is very abundant on the human skin. Singlet oxygen is also one of the components of photochemical smog.

The other toxic forms of oxygen are formed during the reduction of O_2 to H_2O, for example during aerobic respiration for which addition of four electrons is needed. The reactions are shown in Figure 6.4.

$$O_2 + e^- \rightarrow O_2^- \quad \text{superoxide}$$

$$O_2^- + e^- + 2H^+ \rightarrow H_2O_2 \quad \text{hydrogen peroxide}$$

$$H_2O_2 + e^- + H^+ \rightarrow H_2O + OH^{\cdot} \quad \text{hydroxyl radical}$$

$$OH^{\cdot} + e^- + H^+ \rightarrow H_2O \quad \text{water}$$

$$\text{overall:} \ O_2 + 4e^- + 4H^+ \rightarrow 2H_2O$$

Figure 6.4 The reduction of oxygen to water.

superoxide anion

The electrons are usually added stepwise, one electron at a time. The first reduction product is superoxide anion, O_2-. Electron carriers such as flavoproteins, quinones and iron-sulphur proteins can transfer one electron to oxygen thus producing superoxide. Like singlet oxygen, superoxide is very reactive and also oxidises lipids and other important cell components. Of all the oxygen intermediates, which are all short-lived, superoxide has the longest lifespan.

In aerobes, facultative anaerobes and aerotolerant anaerobes accumulation of superoxide is prevented by the enzyme superoxide dismutase, which catalyses its conversion back to oxygen and hydrogen peroxide.

$$2O_2^- + 2H^+ \rightarrow O_2 + H_2O_2$$
$$\text{superoxide}$$
$$\text{dismutase}$$

hydrogen peroxide

The next reduction step leads to formation of peroxide, $O_2$2-, of which hydrogen peroxide is probably the best known form. Hydrogen peroxide is a relatively stable form of peroxide and is a common product of oxygen reduction in respiration, mediated by flavoproteins. It is probably produced by all organisms growing in aerobic conditions.

Most aerobes and aerotolerant anaerobes also posses the enzyme catalase which splits hydrogen peroxide into oxygen and water, thus preventing peroxide accumulation:

$$2H_2O_2 \rightarrow 2H_2O + O_2$$
catalase

The facultative anaerobic group of lactic acid bacteria lacks this enzyme. Still, they do not accumulate significant quantities of hydrogen peroxide. Some of them have peroxidases that catalyse the oxidation of organic compounds by H_2O_2, which is then reduced in water.

hydroxyl free radical
The next reduction product is the hydroxyl free radical, (OH·) the most reactive of all oxygen intermediates. It can affect all cellular components. It is also formed by ionising irradiation, such as X-rays and γ rays, where it is considered to be the principal killing agent.

The enzymes mentioned above (superoxide dismutase, catalase and peroxidase) all protect the cell from toxic effects by oxygen reduction products. Superoxidase dismutase is always present in organisms that can tolerate oxygen, but catalase is not necessarily always there. Neither of these two enzymes, however, has so far been demonstrated in strict anaerobes (Table 6.9).

Bacterium	Superoxide Dismutase	Catalase
Aerobes or facultative anaerobes		
Escherichia coli	+	+
Pseudomonas spp.	+	+
Deinococcus radiodurans	+	+
Aerotolerant bacteria		
Butyribacterium rettgeri	+	-
Streptococcus faecalis	+	-
Streptococcus lactis	+	-
Strict anaerobes		
Clostridium pasteurianum	-	-
Clostridium acetobutylicum	-	-

Table 6.9 Superoxide dismutase and catalase in micro-organisms. + = present, - = not present.

∏ Complete the reactions catalysed by the enzymes shown below.

$$2H_2O_2 \xrightarrow{\text{catalase}}$$

$$2O_2^- + 2H^+ \xrightarrow{\text{superoxide dismutase}}$$

$$H_2O_2 + OCl^- \xrightarrow{\text{peroxidase}}$$

Many of the toxic species of oxygen described above do have specialised functions in some biological systems. These erstwhile toxic oxygen species play an important role in the biodegradation of lignin, polycyclic and aromatic hydrocarbons.

Now check your reactions with those shown in this Section 6.5.2 of text.

6.5.3 Oxygen-sensitive enzymes

Whereas aerobes and aerotolerant micro-organisms have enzymes that protect the cell from toxic effects by oxygen, strict anaerobes have many enzymes that are destroyed by oxygen. A well-known example is nitrogenase, which catalyses nitrogen fixation:

$$N_2 + 8H^+ + 8e^- \rightarrow 2NH_3 + H_2$$

Nitrogenase is inactivated by molecular oxygen

Not only nitrogenase from anaerobes but also those from obligately aerobic nitrogen-fixing bacteria, for instance *Azotobacter*, are extremely sensitive to oxygen when outside (after extraction) the cell. Apparently, when inside the cell, the enzyme is protected from inactivation by special mechanisms, such as high cellular O_2 consumption.

6.5.4 Oxygenases

oxygenases add oxygen atoms to chemicals

Apart from being an electron acceptor in aerobic respiration, oxygen can also act as cosubstrate for enzymes catalysing certain steps in the dissimilation of hydrocarbons. These so-called oxygenases add either one or two oxygen atoms to the organic substrate. Many facultatively anaerobic pseudomonads can use aromatic compounds or alkanes as sole source of carbon and energy. Growth on these substrates in anaerobic conditions (using nitrate as electron acceptor) is only possible when the organism has dehydrogenases able to oxidise the substrate at its disposal, since oxygenases cannot function without oxygen.

Oxygenases are also involved in certain steps in the biosynthesis of sterols and unsaturated fatty acids in eukaryotes and some prokaryotes. These compounds are components of cell membranes. Yeasts need sterols and unsaturated fatty acids when growing fermentatively in anaerobic conditions. When growing aerobically they can make these compounds themselves.

SAQ 6.6

1) Name each of the following forms of oxygen and list in decreasing order of reactivity.

   O_2^{2-}, O_2^-, $^3O_2$, $^1O_2$

2) Four bacterial isolates were shown to possess the following properties.

   Isolate 1: Superoxide dismutase absent.

   Isolate 2: Oxygen not required for growth and catalase present.

   Isolate 3: Oxygen not required for growth and catalase absent.

   Isolate 4; Oxygen required for growth.

   Assign each isolate to the following groups according to their relation to oxygen: facultative anaerobe; aerotolerant anaerobe; obligate aerobe, microaerophile, obligate anaerobe.

6.6 Hydrostatic pressure

All organisms living at the greater depths of deep seas have to endure not only low temperature and marine salt concentrations, but also high external pressure. So micro-organisms living there will certainly be psychrophilic and halophilic, but also be *barophilic*, or, at least, *barotolerant*. (Note, organisms from oceanic vents may be thermophilic and barophilic as well).

barophilic and barotolerant organisms

To find out whether these micro-organisms have a special mechanism to withstand pressures of up to several hundred atmospheres, samples were taken from various depths and the bacteria cultured at corresponding pressure (for every 10 m depth the hydrostatic pressure increases by one atmosphere). The results showed that these deep-sea micro-organisms can be divided into three groups. *Barotolerant* organisms grow best at 1 atm but can adapt to pressures of up to 500 atm. None of the isolated barotolerant bacteria, however, survived pressures of 500 atm and higher. Furthermore it was found that bacteria in samples taken from about 500 m deep grew optimally at about 400-500 atm, which indicates that they are moderate *barophiles*, although they could also grow at 1 atm.

Bacteria in samples from much greater depths, 10,000 m (1000 atm!) proved to be *extreme or obligate barophiles* and could not grow at atmospheric pressure. Figure 6.5 shows how bacteria can be divided according to their tolerance for hydrostatic pressure.

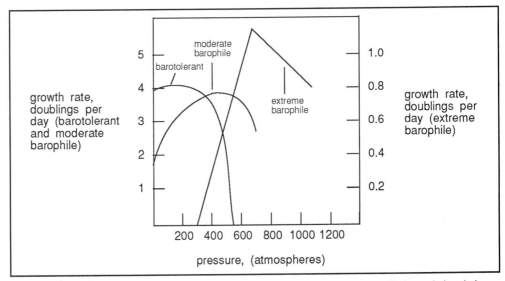

Figure 6.5 Growth rate of barotolerant, moderately barophilic and extremely barophilic bacteria in relation to hydrostatic pressure.

effects of hydrostatic pressure on the cell

One effect of hydrostatic pressure on the microbial cell is the decrease in binding-capacity between enzymes and their substrate and hence in enzyme activity. However, not all enzymes are inhibited at the same rate. It is, for instance, known that respiration processes are less affected than protein synthesis. Also several membrane bound processes, such as transport, are influenced by high pressures.

SAQ 6.7

1) According to their relation to hydrostatic pressure, bacteria are grouped as barotolerant, moderate barophiles or extreme barophiles.

   To which, if any, of these groups would you expect organisms described as a) facultative barophiles and b) obligate barophiles to belong? Give a reason for your response in each case.

2) In general would you expect the growth rates of barophiles to be higher or lower than that of non-barophiles? Give a reason for your response.

3) A bacterium isolated from a depth of 10,000 m grew well at 600-800 atmospheres but not at all at 1000 atmospheres. Group the isolate according to its relation to hydrostatic pressure.

6.7 Radiation

Several forms of radiation can cause a damaging effect on micro-organisms. In this section we will examine the effect of visible light, ultraviolet and ionising radiation respectively. The wavelengths of these types of radiation are shown in Figure 6.6.

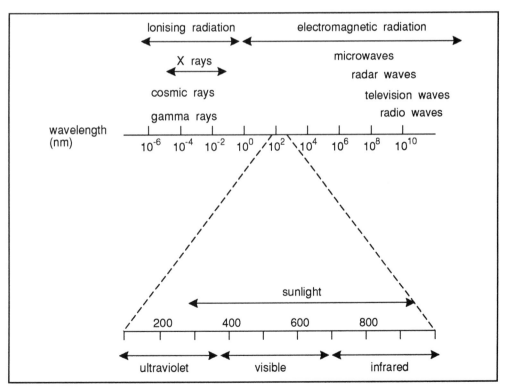

Figure 6.6 Wavelengths of radiation.

6.7.1 Visible light

When sufficiently intensive, visible light can cause damage to the microbial cell and can even cause death. Most organisms have cell components that are sensitive to light. The light-absorbing cell components (cytochromes, flavins and chlorophylls) become activated when absorbing light and are raised to a higher energy state. They can then either return to their ground state by emitting light (fluorescence) or by transferring energy to another cell component. The energy transfer may be advantageous to the organism (photosynthesis) or damaging. In the latter case there are two mechanisms which can cause the harmful effect, one of these involves molecular oxygen. The oxygen-independent damage is caused by the formation of free radicals, which are extremely reactive and destructive. Much more damage, however, is caused by the oxygen-dependent mechanism in which singlet oxygen is involved, this has already been discussed in Section 6.5.

visible light generates free radicals and singlet oxygen

∏ Explain why a combination of aerobic growth in light on white (unpigmented) *M. luteus* is different to the other combinations shown in Figure 6.7.

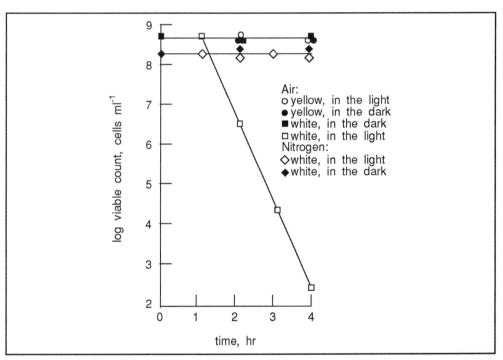

Figure 6.7 Effect of sunlight on the viability of yellow- pigmented or white (non-pigmented) strains of *Micrococcus luteus*, under anaerobic (nitrogen atmosphere) and under aerobic conditions. Data from Brock, T.D, Smith, D.W and Madigan, MT (1984) Biology of Micro-organisms, Prentice Hall.

The destructive effect of this combination is due to a combination of the deleterious effect of strong light on unprotected (non-pigmented) cells and the availability of oxygen from air to produce singlet oxygen or oxygen radicals.

role for carotenoids

We have already seen (Section 6.5) that some of the pigments present in microbial cells protect the cell from harmful effects by singlet oxygen and, of these, the carotenoids are the most effective. Interaction of the carotenoid and singlet oxygen changes the spin of the oxygen molecule, which brings singlet oxygen back to its ground (triplet) state. By this the carotenoid becomes activated and returns to its ground state spontaneously. The protective effect of carotenoid is clearly shown in Figure 6.7. The *Micrococcus* strain lacking carotenoid is killed by light and the one with carotenoid is not affected.

∏   It has been shown that carotenoid-containing bacteria are protected against killing by phagocytes, which is of clinical significance since pathogenic strains of *Staphylococcus* are nearly always pigmented. Can you explain how these bacteria are protected?

In Section 6.5.2 we mentioned that singlet oxygen is formed in phagocytic animal cells, where it can kill ingested micro-organisms. It is clear that micro-organisms containing carotenoids will not be affected by this defence mechanism of the body.

6.7.2 Ultraviolet radiation

Of the ultraviolet (UV) radiation emitted by the sun, only that of the longer less harmful wavelengths reaches the earth's surface. However shorter wavelength UV-light is an important natural killing agent of micro-organisms and is applied in a number of 'germicidal lamps'.

UV-light damages DNA
Of the cellular components DNA is most severely affected by UV radiation and damage to DNA is the main cause of cell death by UV irradiation. Purine and pyrimidine bases of both DNA and RNA absorb UV light strongly, with a peak at 260 nm. Proteins are also affected. They absorb UV light at 280 nm, which is where the aromatic amino acids (tryptophan, phenylalanine, tyrosine) have their absorption peak.

One of the effects of UV radiation on DNA is the formation of a covalent bond between two adjacent thymine molecules. The so-formed thymine dimers prevent replication of DNA. Most micro-organisms have enzymes capable of repairing the induced damage to the DNA molecule which is usually in one strand only.

repair enzymes
There are, for instance, repair enzymes that can hydrolyse the phosphodiester bonds at each end of the damaged region, so that the altered material (with the thymine dimer) in it, is excised. Next, insertion of new nucleotides, complementary to the intact strand, is then catalysed by other repair enzymes. There is one kind of repair enzyme that only does its repairing job when the damaged cell is exposed to visible light (of the blue region of the spectrum), so-called photoreactivation (see Figure 6.8).

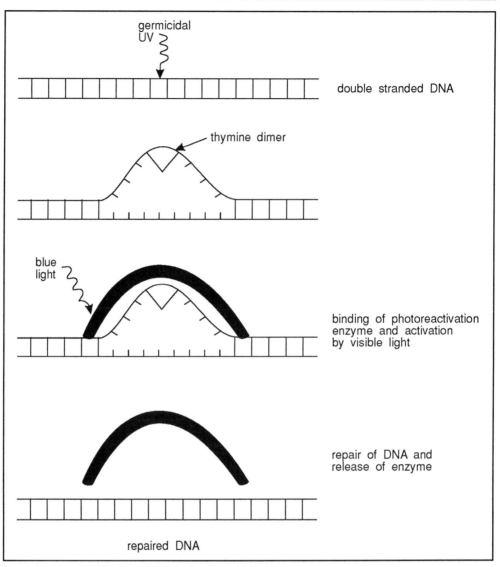

Figure 6.8 Photoreactivation of UV-light damage to DNA.

Other repair enzymes work in the dark: dark reactivation. Whether a cell is fatally damaged depends on the degree of damage and the repair capacity of the enzymes.

∏ Damage to DNA is likely to have more serious consequences than damage to proteins. Try to explain.

There is usually only one copy of DNA present in the cell whereas there are usually many copies of protein molecules. The damage to DNA persists in subsequent generations.

UV-light is an efficient form of sterilisation but is generally limited to gases as it is non-penetrative. For instance a few sheets of paper will prevent UV penetration.

6.7.3 Ionising radiation

ionising
radiation forms
free radicals

Of the various forms of radiation, ionising radiation is a killer. Unlike UV radiation it does not affect the cell components directly but works indirectly by the formation of free radicals in the medium. Of these the hydroxyl radical (OH·) is the most important one, being very reactive. Although free radicals can act on most cell components, their lethal effect is mainly caused by damage to DNA. This is not because DNA is more sensitive than any of the other macromolecules in the cell but because there is only one copy of most genes. So damage to one gene is more devastating than damage to, for instance one single protein molecule of which there are likely to be many copies. Damage by ionising radiation can also be repaired, but only by dark reactivation.

An application of the killing capacity of ionising radiation is sterilisation of products (food, drugs) and materials that cannot be sterilised by heat which is the more commonly used method for sterilising purposes. An advantage of this way of sterilising is that ionising radiation can penetrate many materials, so that sterilisation can take place after packaging.

Ionising radiation can be generated by X-ray apparatus and radioisotopes. For sterilisation the radioisotope cobalt-60 is the most convenient radiation source.

Since ionising radiation has a great penetration capacity and can kill human cells as well, working with this kind of radiation requires extensive precautions, such as protection by a lead shield and careful monitoring of the operators.

Π The lethal dose of ionising radiation for prokaryotes is about 100-1000 times as high as that for a mammalian cell. Can you explain why this is so?

Statistically, the chances for a lethal package of energy and a macromolecule to hit each other is proportional to the size of the latter. The observed difference in sensitivity can, to a large extent, be explained by the fact that the DNA molecule in eukaryotes is much larger and more complex than in prokaryotes. The available repair mechanisms also play a part.

SAQ 6.8

For each of the following types of radiation, select appropriate killing mechanism(s), wavelength range and protection mechanism(s) from the lists provided.

Radiation	Killing mechanism
1) UV	None
2) Visible light	Free radicals
3) X-rays	Dimer formation in DNA.
4) Radiowaves	Singlet oxygen
Wavelength range (nm)	**Protection mechanism**
400-700	The atmosphere
$10^{-4} \rightarrow 10^{-2}$	Carotenoids
~ 10^{10}	Photoreactivation
100-400	Dark reactivation
	None

SAQ 6.9

A new isolate exhibits all of the properties listed 1) to 7).

1) Requires 6% NaCl to grow.

2) Grows at a water activity of 0.7.

3) Its optimum growth temperature is 25°C but it can grow at 0°C.

4) Can only be cultivated on high sugar containing medium.

5) Grows well at pH 2.0 and not at all at pH 7.0.

6) Oxygen is required for growth but it does not grow optimally in air or in an oxygen enriched atmosphere.

7) Grows at an atmospheric pressure of 400 but not at 600 atmospheres.

Use appropriate terms to describe the organism as fully as possible.

Summary and objectives

In this chapter we have seen that micro-organisms can be categorised according to their relation to temperature, water availability, pH, hydrostatic pressure and oxygen. Most of the categories are defined by the minimum, maximum and optimum values for growth. Many micro-organisms have special characteristics/processes that protect them from environmental extremes, these include: altering the fatty acid composition of the cell membrane (temperature); increasing osmolarity of the cytoplasm (water availability); a cell membrane with relatively low permeability to cations (pH); special enzymes (oxygen and radiation); pigments (oxygen and radiation).

Oxygen has different electronic configurations and some are highly reactive with biological macromolecules. The toxic effect of visible light and X-rays is also caused by the generation of highly reactive forms of oxygen. In the case of UV light the killing mechanism is thymine dimer formation in DNA.

Now that you have completed this chapter you should be able to:

- define categories of micro-organisms according to their relation to temperature, water availability, pH, oxygen and hydrostatic pressure;

- give named examples of micro-organisms for each category;

- describe the effects of environmental extremes on micro-organisms;

- list characteristics and describe processes that protect micro-organisms from adverse effects of temperature, water availability, pH, oxygen and radiation;

- list toxic forms of oxygen and describe their electronic configuration;

- distinguish various types of radiation according to wavelength and their interaction with cellular macromolecules.

Control of environmental factors influencing growth

Control of environmental factors influencing growth

In the preceding chapter we saw that there are several environmental factors that influence the growth of micro-organisms. It follows that to maximise growth and/or product formation *in vitro*, these environmental factors should be maintained at desired values. This often requires three different functions: measurement, analysis of measurement data, and control action.

In this chapter we will consider the various strategies for the measurement and control of the main environmental factors influencing growth processes in fermentors; these are pH, availability of oxygen, and temperature. Here we use the term of fermentor to describe a bioreactor in which we cultivate cells. Do not confuse this term with fermentation. Strictly, fermentation is the term applied to a form of metabolism that does not require molecular oxygen.

7.1 pH

We saw in Chapter 6 that pH is a particularly important factor affecting microbial growth. You will recall that the pH is inversely related to the hydrogen ion concentration, that is:

$$pH = \log \frac{1}{[H^+]}$$

It is obvious why culture media should be prepared at a particular pH. However, once the pH is set there may still be a requirement for the control of pH.

∏ Can you think of two reasons why pH control may be necessary during cultivation of micro-organisms?

There are two fundamental reasons:

- growth of micro-organisms tends to change the pH of their environment;

- it may be desirable to control a fermentation by altering the pH of the medium during growth, or after growth has finished, for example to prevent breakdown of a product formed during a fermentation process.

7.1.1 Changes in pH as a result of microbial activity

The pH of a metabolizing culture will not remain constant for long. These changes of pH are associated with:

- the degradation of proteins and other nitrogenous compounds with the formation of ammonia or other alkaline products;

- the uptake of certain anions and cations;

- the metabolism of carbon substrates with the formation of organic acids.

With some specialist groups of micro-organisms we could add:

- the use of nitrate or sulphate as terminal electron acceptors;

- the use of NH_4^+ or sulphate as an energy source.

We will now consider some examples, which illustrate how medium composition influences pH changes during growth.

pH effects of using ammonium salts

Example 1: In culture media ammonium salts, such as $(NH_4)_2SO_4$, are often used as a source of nitrogen for growth. Utilisation of ammonia from these sources leaves in the medium a corresponding amount of free acid, as follows:

$$(NH_4)_2SO_4 \longrightarrow 2NH_3 + H_2SO_4 \text{ (acid)}$$

$$\searrow$$

biomass

pH effects of using nitrates

Example 2: Nitrate salts, such as KNO_3, can also be included in culture media as a source of nitrogen for growth. In these media, the utilisation of nitrate consumes H^+ and thus tends to increase pH, as follows:

$$KNO_3 + H_2O \longrightarrow NO_3^- + KOH \text{ (alkali)}$$

$$H^+ \nearrow$$
$$e^- \nearrow$$

$$NH_3 \longrightarrow \text{biomass}$$

pH effects of using amino acids

Example 3: An alternative to the provision of an inorganic source of nitrogen is the supply of amino acids. These are added to culture media as protein hydrolysates and can serve as a combined nitrogen source and major or sole carbon source. In these media, the aerobic metabolism of amino acids leads to the production of ammonia in excess to the cells nitrogen requirement. The ammonia production will result in OH^- ions and an increase in pH, as follows:

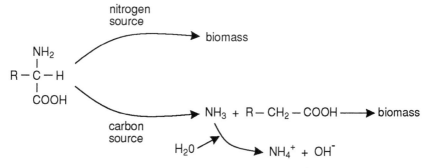

The anaerobic metabolism of amino acids is more complex with respect to pH: decarboxylase activity will result in alkaline amine products, whereas deaminase activity will give both acid and alkali products.

pH effects of using carbohydrates

Example 4: If the major carbon source in a culture medium is a carbohydrate, as indeed it often is, the pH may drop. The explanation for this depends on whether the organism respires or ferments the sugar. During exponential growth under aerobic conditions, many organisms produce partially metabolised compounds such as acetic and pyruvic acid in the presence of excess sugar.

During anaerobic growth of carbohydrates acidic or neutral products are produced. The range of these products depends on the organism and the conditions of culture. For example, *Escherichia coli* is a mixed acid fermenter and metabolises glucose as follows:

glucose → lactic acid + acetic acid + ethanol + CO_2 + H_2

pH effects of using CO_2

Example 5: In the case of autotrophic organisms, such as algae, the major carbon source is carbon dioxide. The level of available CO_2 can be boosted by the addition of bicarbonate salts, such as $NaHCO_3$, to culture media. Uptake of CO_2 from bicarbonate tends to increase pH by mopping up H^+ as follows:

$$HCO_3^- \xrightleftharpoons[]{H^+} H_2CO_3 \rightleftharpoons CO_2 + H_2O$$

bicarbonate   hydrogen
                                carbonate

SAQ 7.1

For each of the following batch cultivations, predict whether the pH will decrease or increase and give a reason for your decision (assume that the medium has only low buffering capacity).

1) Growth of autotrophic algal culture where the level of CO_2 is boosted by incubating the culture in a CO_2 enriched atmosphere.

2) Growth of an aerobic bacterial culture where a protein hydrolysate is provided as a combined carbon and nitrogen source

3) Growth of an anaerobic bacterial culture where a protein hydrolysate is provided as nitrogen source and glucose is provided as major carbon source.

4) Growth of an aerobic bacterial culture where gluconic acid is provided as sole source of carbon and energy.

5) Growth in a culture medium containing nitrate as source of nitrogen.

7.1.2 Control of pH

An obvious way of preventing a pH change during microbial growth is to incorporate a buffer into the culture medium.

∏   Buffers for culture media should have particular characteristics eg they should be water soluble. Can you list some additional desirable characteristics?

Now compare your list with the following six main desirable characteristic of culture media:

• the buffer should be very soluble in water and non-volatile;

- the medium concentration, temperature and ionic strength should have a minimum effect on the capacity of the buffer;

- the buffer should not absorb light in the visible or ultraviolet regions of the spectrum (this would interfere with spectrophotometric measurement of growth or product formation);

- the components (protonated and non-protonated forms) of the buffer should not be toxic to the micro-organism or have any inhibitory side effects;

- in order to minimise unwanted effects on the organism, the buffer should not readily pass through biological membranes;

- the buffer should be non-metabolizable and resist enzymatic degradation.

For culture media, phosphate buffers are often used. There are two main reasons for this:

- phosphate buffers are effective over a pH range (6.5 to 8.0) that includes the optimum pH for most micro-organisms;

- they are usually non-metabolizable and do not adversely affect the growth of most micro-organisms.

We remind you of the Henderson-Hasselbach equation which relates pH to the pKa and the concentration of a weak acid and its conjugate base. K_a is the dissociation constant of the acid.

Thus for the acid HA which disociates thus:

$$HA \xrightleftharpoons{} H^+ + A^-$$

$$pH = pKa + \log \frac{[A^-]}{[HA]}$$ where [ ] signifies concentration.

Weak acids buffer best at pHs close to their pKa values (If you are unfamiliar with these terms, we recommend the BIOTOL text 'The Molecular Fabric of Cells'). Phosphoric acid has three dissociations thus:

$$\underset{H_3PO_4}{\overset{pKa = 2.1}{\xrightleftharpoons{}}} \quad H^+ + H_2PO_4^- \quad \overset{pKa = 7.2}{\xrightleftharpoons{}} \quad H^+ + HPO_4^{2-} \quad \overset{pKa = 12.3}{\xrightleftharpoons{}} \quad H^+ + PO_4^{3-}$$

Usually phosphate buffers are used at pH's around 6.5 to 8.0 and thus involved the ions $H_2PO_4^-$ and $H_2PO_4^{2-}$.

Potassium phosphate buffers, for example, are composed of K_2HPO_4 (a weak alkali) and KH_2PO_4 (a weak acid). Equimolar concentrations of these components give a solution with a pH of 7.2. The HPO_4^{2-} component resists a decrease in pH by converting a strong acid to a weak acid:

$$HPO_4^{2-} + HCl \longrightarrow H_2PO_4^- + Cl$$
$$\phantom{HPO_4^{2-} + }\text{strong} \qquad \text{weak}$$
$$\phantom{HPO_4^{2-} + }\text{acid} \qquad \text{acid}$$

Conversely, the $H_2PO_4^-$ component resists an increase in pH by converting a strong alkali to a weak alkali:

$$H_2PO_4^- + KOH \longrightarrow HPO_4^{2-} + H_2O$$
$$\text{(strong alkali)} \quad \text{(weak alkali)}$$

reserve alkali additions

Where a large amount of acid is produced through growth, the buffering capacity of phosphates might not be sufficient. If this is expected then it is sometimes appropriate to add a reserve of alkali to the culture medium. This is usually a neutral carbonate, such as calcium carbonate ($CaCO_3$) added as finely powdered chalk.

Π Identify the products at position a), b) and c) in the reaction sequence shown below. Explain how carbonate acts as a reserve of alkali in culture media.

$$H^+ + CO_3^{2-} \longrightarrow a) \overset{H^+}{\longrightarrow} b) \longrightarrow c)$$

The products are: a) bicarbonate (HCO_3^-), b) hydrogen carbonate (H_2CO_3) and c) CO_2 and H_2O. From this scheme you can see that carbonate acts as a reserve of alkali by mopping up H^+ ions.

Calcium carbonate is used because it is relatively insoluble. Thus CO_3^{2-} will be released slowly into the medium at a rate dependent on its requirement to maintain pH.

Similarly, in fermentation processes the buffering of culture media is usually not adequate for the control of pH, since high biomass concentrations are usually desirable. Here, it is usual to continually monitor pH, and control by automatic addition of a suitable acid or alkali. A limitation of this approach is that the additions may adversely alter the chemical environment of the cells. Acid production by the culture can be titrated using NaOH or $Ca(OH)_2$. However, these will affect the ionic strength of the medium and may also affect product solubility. In the case of H^+ generation from ammonia utilisation by the culture, a constant pH and NH_4^+ concentration can be maintained by titrating with ammonia. However, if ammonia is used to titrate an organic acid that is being formed, the culture may develop ammonia toxicity. In the case of an alkali swing, it is usual to titrate with hydrochloric (HCl) or sulphuric acids (H_2SO_4).

toxity problems

Since most fermentations produce acid, the controller usually only needs to govern the additions of alkali. Despite this, however, it is often the practice to set up both acid and alkali feeds to the vessel. In the case of an acid-producing fermentation, alkali is normally added, but if there is an overdose of alkali, the controller will correct this by adding acid.

With an automatic pH control system the volume of acid/alkali added is an important consideration. In batch culture this will influence the final volume of the culture and in continuous culture it may influence the steady-state dilution rate.

Π Can you think of some factors that would govern the volume of acid/alkali required for pH control of a fermentation?

There are three main factors:

- the rate of production of acid/alkali by the culture;
- the strength of the acid/alkali being added;
- the type of computer control used.

Whereas the first two factors are quite obvious, the third requires explanation and will be considered later in this chapter (Section 7.5). You may have included buffering capacity (strength) of the medium in your list of factors, however, we shall see that although this is an important consideration it does not influence the volume of control reagent added.

The concentration of the acid/alkali added influences the rate of addition as well as the volume of reagent added. With a concentrated solution the mixing time of the fermenter and the response time of the pH sensor may be too slow, resulting in 'overshoot' and large pH oscillations. A dilute solution may result in large volume additions and, in continuous culture, a significant contribution to dilution rate. Table 7.1 shows how medium buffering capacity and concentration of acid/alkali reagents influence pH control for a continuous culture.

Buffer molarity	mmole equivalents of alkali required to change pH from 6.5-6.7	Number of additions/hour of concentrated alkali (moles l^{-1})		
		1	2	4
0.01	0.6	65	33	16
0.025	1.5	27	14	7
0.05	3	13	6	3
0.1	6.1	6.5	3.5	1.7
0.2	12.2	3.2	1.6	0.8
Volume of alkali added		40 ml	20 ml	10 ml
Volume of alkali added as % of dilution rate		20	10	5

Table 7.1 Automatic pH control of a continuous fermentation. The right hand column shows the number of additions that need to be made with alkali of different concentrations. Thus 65 additions of 1 mol alkali l^{-1} are required when the buffer molarity is 0.01. On the other hand, if 4 mol alkali l^{-1} is used, only 16 additions need to be made (see text for discussion).

We can see from Table 7.1 that the buffering capacity of the medium influences the number of additions required per hour but does not alter the volume of acid/alkali required for pH control. Since the buffering capacity of the medium determines the ability to neutralise the control reagent without great change in pH, for automatic control the buffer concentration should be minimal. The data shown in Table 7.1 can be used to decide what buffering capacity and strength of control reagents are compatible with a chosen dilution rate. We can see from the data that when 2 moles l^{-1} alkali is used the dilution rate will be 10% more than that determined by measuring medium flow rate alone. It follows that the volume of acid/alkali added must be known to successfully calculate the dilution rate. In practice, addition of reagents whose volume is less than 1-2% of the medium flow rate can probably be neglected in most applications, especially since the accuracy of measuring the medium flow rate may be no greater than 1-2%.

SAQ 7.2

1) During automatic control of pH by acid/alkali addition, how will each of the following (i to iii) be altered by a) a decrease in the molarity of control reagent (acid/alkali) and b) an increase in medium buffer molarity.

   i) The number of additions per hour.

   ii) The volume of acid/alkali added per hour.

   iii) The moles equivalent of acid/alkali required to change pH.

2) Identify each of the following statements as true or false. If false give a reason for your response.

   a) Citric acid buffer is widely used in culture media.

   b) K_2HPO_4 resists an increase in pH.

   c) Phosphate buffers control pH over the range 5.0 - 7.5.

   d) A buffer should, ideally, pass through biological membranes.

7 2 Availability of oxygen

During growth of aerobic micro-organisms in fermentors, the provision of sufficient oxygen for metabolic purposes is an important consideration and often a problem. The problem arises because many processes are operated at high biomass concentrations, so oxygen demands by these cultures are high. Since the solubility of oxygen in culture media is a small fraction of that required by the cultures, oxygen used must be replaced continuously and efficiently. How this can be achieved is considered in this section. Provision of sufficient oxygen for microbial growth has been the subject of intense research for many years and continues to be an area of interest. Lets firstly examine the problem: high oxygen demand versus low oxygen solubility.

7.2.1 Microbial oxygen demand

In Chapter 6 we saw that oxygen serves as the final electron acceptor in aerobic metabolism or may be incorporated into carbon substrate via oxygenase enzymes. For example, glucose oxidation during growth of *Escherichia coli* can be described as follows:

$$C_6H_{12}O_6 + 6O_2 \rightarrow 6CO_2 + 6H_2O$$

When hydrocarbons are used as the energy source then oxygen is required not only to produce energy but also for raising the oxidation level of substrate to that of the biomass. It follows that production of biomass from a hydrocarbon requires more oxygen than production from carbohydrate. The amount of oxygen required can be calculated from the following balance:

Oxygen required to produce 1g dry biomass (C) = oxygen required to oxidise substrate used to produce 1g dry biomass - oxygen required to oxidise 1g dry biomass (B).

This is expressed as:

$$C = A/Y_s - B \qquad\qquad (E - 7.1)$$

where:

A = amount of oxygen required for the oxidation of 1g substrate to CO_2 and H_2O (and NH_3 if the substrate contains nitrogen).

Y_s = yield coefficient for carbon substrate (g g^{-1})

∏ Yield coefficients for glucose ($C_6H_{12}O_6$) and for methane (CH_4) were found to be 0.51 and 1.03 respectively. If availability of oxygen limits the biomass output of the fermenter, by what factor does the maximum output rate with glucose as substrate exceed that with the n-alkane (Oxygen required to oxidise 1g dry biomass = 41.7 mmol oxygen g^{-1} dry weight).

For glucose (molecular weight = 180)

$$C_6H_{12}O_6 + 6O_2 \rightarrow 6\ CO_2 + 6H_2O$$

using equation 7.1: $C = A/Y_s - B$

1g substrate = $\dfrac{1}{180}$ mol substrate and is oxidised by $\dfrac{6}{180}$ mol O_2 = 33.33 mmol O_2 g^{-1} substrate

Thus $C = \dfrac{33.33}{0.51} - 41.7 = 23.6$ mmoles g^{-1} biomass

For methane (molecular weight = 16)

$$CH_4 + 2O_2 \rightarrow CO_2 + 2H_2O$$

Thus 1g substrate = $\dfrac{1}{16}$ mol substrate which is oxidised by $\dfrac{2}{16}$ mol O_2 = 125 mmol O_2 g^{-1} substrate.

Thus $C = \dfrac{125}{1.03} - 41.7 = 79.6$ mmoles g^{-1} biomass.

It follows that maximum output rate with glucose as substrate will exceed that with methane by a factor of: 79.6/23.6 = 3.4

The approach determines the total amount of oxygen required but does not provide any information about when during cultivation the oxygen is needed. This is given by the specific O_2 uptake rate (q_o), and typically has units: mmoles O_2 g^{-1} dry weight biomass h^{-1}. Table 7.2 gives examples of maximum specific uptake rates values.

Organism	$q_o$max $(m \text{ moles } O_2 \text{ g}^{-1}\text{h}^{-1})$
Aspergillus niger	3.0
Streptomyces griseus	3.0
Penicillium chrysogenum	3.9
Klebsiella aerogenes	4.0
Saccharomyces cerevisiae	8.0
Escherichia coli	10.8

Table 7.2 Maximum specific uptake rate of some micro-organisms.

oxygen transfer rate (OTR)

The value of q_o is dependent upon the specific growth rate (μ) and the biomass yield coefficient for oxygen (Y_o): $q_o = \mu/Y_o$. When q_o is correlated with biomass concentration (X), we obtain the oxygen uptake rate or oxygen transfer rate (OTR) for the culture:

$$OTR = q_o.X$$

So,

$$OTR = (\mu/Y_o).X \qquad\qquad (E - 7.2)$$

We saw in Chapter 3 that this relationship (equation 7.2) could be used to determine the biomass concentration. It can also be used to determine the OTR, provided of course, the specific growth rate (μ), the biomass concentration (X) and the yield coefficient for oxygen (Y_o) are known.

∏ Determine the OTR in suitable units using the following information. $\mu = 0.2 \text{ h}^{-1}$. $Y_o = 0.5$ g biomass g^{-1} oxygen. X = 5.0 g l^{-1}.

Using equation 7.2:

$$OTR = \frac{0.2 \text{ h}^{-1}}{0.5 \text{ g biomass g}^{-1} \text{ oxygen}} \, 5.0 \text{ g biomass l}^{-1}$$

$OTR = 2.0$ g oxygen l^{-1} h^{-1}.

You will recall from Chapter 3 that the value of Y_o can be determined from the growth yield coefficient (Y_s) for carbon substrate, taking into account the elemental composition of biomass and of carbon substrate.

∏ Explain what happens to the OTR during batch growth.

changes in OTR in batch cultures

The OTR increases during exponential phase because the biomass concentration is increasing. This continues until the culture becomes limited by the OTR or by another nutrient. In the latter case, the OTR then falls since the specific growth rate (μ) is lowered.

Together with the OTR, the concentration of oxygen in culture media is an important consideration. Indeed, we shall see later in this chapter that dissolved oxygen concentration directly influences OTR. In addition, the dissolved oxygen concentration must be above a minimum value to permit respiration without hinderance. This minimum value is termed the critical oxygen concentration and is typically 5-20% of the oxygen saturation value for aqueous media.

The solubility of oxygen in aqueous media in equilibrium with air at one atmosphere is, in fact, only a few mg l^{-1}. We can see from Table 7.2 that this is slight compared with the amount of oxygen that can be consumed by a culture.

∏ The oxygen concentration in water at 25°C is 8.1 mg l^{-1}. Assuming 100% utilisation of oxygen, what is the shortest possible time for the oxygen to be removed by an *E. coli* culture at a biomass concentration of 10 g l^{-1}? (Use the data in Table 7.2)

From Table 7.2 the maximum rate of oxygen removal by *E. coli* is 0.345 g O_2 g^{-1} biomass h^{-1}. So, the oxygen demand by the culture is: 0.345 x 10 = 3.45 g oxygen $l^{-1}h^{-1}$ or 0.958 mg l^{-1} s^{-1}. This demand would remove 8.1 mg of oxygen in: $\dfrac{8.1}{0.958}$ = 8.35 seconds.

In the example, the actively respiring population consumes oxygen at a rate which is of the order of 430 times the saturation value per hour. So, how can the high oxygen demand of microbial cultures be satisfied? One approach to increasing the availability of oxygen to microbial cultures is to increase the dissolved oxygen concentration. To establish whether this is a realistic approach to satisfying microbial oxygen demand, lets consider the main factors which affect solubility of oxygen. These are:

- partial pressure of oxygen;

- temperature;

- other solutes in the medium.

The effect of oxygen pressure in the gas phase (p_o) is given by Henry's law:

$$C^* = p_o/H \qquad\qquad (E - 7.3)$$

Where:

C^* = dissolved oxygen saturation concentration (mg l^{-1});

H = Henry's constant which is specific for the gas and liquid phase (bar 1 g^{-1} oxygen);

p_o = partial pressure (bar); also known as the oxygen tension or activity.

Henry's law shows that as the oxygen concentration increases in the gas phase the dissolved oxygen concentration increases in aqueous medium. It follows that the highest dissolved oxygen concentrations are obtained during aeration with pure oxygen. Compared with air almost five times more oxygen dissolves in water when pure oxygen is used.

Conversely, oxygen solubility decreases with increase in temperature and with increase in dissolved solutes. This increases the partial pressure of oxygen in the gaseous phase, which means that the activity of oxygen is increased. Table 7.3 shows the influence of temperature and dissolved solutes on the solubility of oxygen.

Temperature $^\circ$C	Oxygen solubility in water (mg l^{-1})	*Na Cl concentration (mol l^{-1})	O_2 solubility (mg l^{-1})
10	10.9	0	8.1
20	8.8	0.5	6.8
30	7.5	1.0	5.6
40	6.6	2.0	4.5

Table 7.3 Influence of temperature and dissolved solutes on the solubility of oxygen. Water in equilibrium with air at 1 bar (atmosphere). (p_o = 0.209 bar) * Aqueous solution at 25°C

∏ Explain how you would expect the addition of a solute to aqueous media to influence Henry's constant?

We know that the addition of solute to aqueous media decreases the dissolved oxygen concentration and increases partial pressure of oxygen. It follows from Henry's law (7.3) that Henry's constant (H) must increase. This emphasises the fact that Henry's constant is specific for a particular gas and liquid phase.

Of the three factors influencing oxygen solubility, the use of pure oxygen is the only realistic possibility for increasing availability of oxygen in cultures. Indeed, this approach has been used for some small scale fermentations. However, the costs associated with the use of pure oxygen makes it unsuitable for industrial processes.

7.2.2 Mass transfer of oxygen

Clearly, the factors that influence solubility of oxygen will not satisfy the high oxygen demand of fermentations. It follows that to ensure continued growth of microbial cultures, the dissolved oxygen removed by cells must be replaced continuously and efficiently by oxygen from the air.

For oxygen to be transferred from a gas bubble to an individual cell for use in metabolism, it must pass through a series of transport resistances. These are illustrated in Figure 7.1.

Microbial cells near gas bubbles may absorb oxygen directly through the gas-liquid boundary and the rate of gas transfer to these cells is faster than to others. For fermentations carried out with unicellular organisms such as bacteria or yeasts, the resistance of the gas-liquid boundary is the most important factor controlling the rate of oxygen transfer. However, in the case of cell flocks and pellets of filamentous growth, the additional resistance of diffusion transport into the cellular mass may be the most important factor.

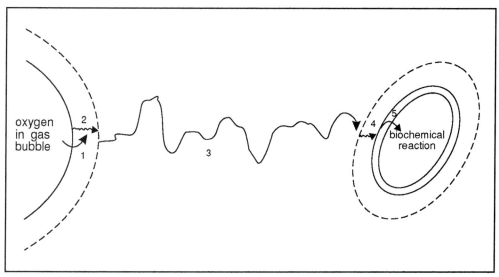

Figure 7.1 Steps involved in transport of oxygen from a gas bubble to inside a cell. 1) movement through the gas-liquid interface. 2) diffusion of the solute through the relatively unmixed liquid region adjacent to the bubble into the well mixed bulk liquid. 3) transport of the solute through the bulk liquid to a second relatively unmixed liquid region surrounding the cell. 4) transport through the second unmixed liquid region surrounding the cell. 5) transport across cell envelope and to the intracellular reaction site.

Π Can you think of two factors that would influence the thickness of the gas-liquid boundary in a fermentation process?

Two of the main factors are viscosity of the medium (high viscosity increases thickness) and agitation of the culture (reduces thickness).

We have already seen that oxygen transfer rate (OTR) can be described according to growth parameters (Equation 7.2). For mass transfer studies, however, it is more useful to describe the OTR in terms of dissolved oxygen concentrations, as follows:

$$OTR = k_L.a \, (C^* - C_L) \qquad\qquad\qquad\qquad (E - 7.4)$$

where:

k_L = mass transfer coefficient at the gas-liquid boundary (m h^{-1});

a = the gas-liquid interfacial area available for mass transfer per unit volume (m^{2}m^{-3} = m^{-1});

k_La = volumetric transfer coefficient;

C^* = oxygen saturation concentration of nutrient medium (usually mmol l^{-1} or mg l^{-1});

C_L = concentration of dissolved gas (units as for C^*)

k_La volumetric oxygen transfer coefficient   In this expression the driving force which moves oxygen across the gas-liquid boundary is the difference in concentrations. Whereas k_L may be considered as the ease with which oxygen is transferred from gas to liquid. The volumetric oxygen transfer coefficient (k_La) is a critical parameter in fermentation process design.

∏ Establish suitable units for k_La and for OTR.

Since k_L has units of $m \, h^{-1}$ and 'a' has units of m^{-1}, it follows that k_{La} has units of h^{-1}. From Equation 7.4 we can see that OTR will have units of $mmol \, l^{-1}h^{-1}$ or $mg \, l^{-1}h^{-1}$.

∏ Can you think why the continual supply of an adequate amount of oxygen for microbial respiration should not be regarded as a trivial task? (Hint: consider Equation 7.4).

The low oxygen solubility guarantees that the concentration difference which is the driving force for oxygen transfer across the boundaries is always very small.

SAQ 7.3

Identify each of the following statements as True or False and give a reason for your choice.

1) Increasing the partial pressure increases the volumetric transfer coefficient.

2) In an aerated culture, the OTR increases with decrease in bubble size.

3) At a constant partial pressure the concentration of oxygen in culture media will generally be lower than that of pure water.

4) Addition of solute to culture media will not influence the value of Henry's constant.

5) OTR increases with increase in biomass yield coefficient for oxygen (Y_o).

6) The driving force for oxygen transfer across a gaseous-liquid boundary is proportional to the interfacial area.

---

determination of k_La

The values of 'k_L' and 'a' are difficult to calculate individually. This is because the interfacial area is usually unknown since it depends on bubble size. However, the combined parameter, k_La, can be measured. It is used as a measure of the aeration capacity of a bioreactor and is a critical parameter in fermentation process design. We will now consider various methods of k_La determination.

Oxygen balance method

This is the most reliable method of k_La determination and involves the measurement of oxygen concentration in input and exit air. The difference in their concentration, when multiplied by the aeration rate, and taking absolute temperature and pressure into consideration gives the oxygen transfer rate (OTR).

$$OTR = Q \, (Y_{in} - Y_{out}) \qquad\qquad (E - 7.5)$$

where:

Q = aeration rate;

Y_{in} = oxygen content of the input;

Y_{out} = oxygen content of the exit.

The definition of oxygen transfer coefficient (Equation 7.4) is then used to determine k_La ie OTR = k_La (C^* - C_L)

For this, the medium saturation concentration (C^*) is determined using Henry's law (Equation 7.3), which relates dissolved oxygen concentration to partial pressure of oxygen. A dissolved oxygen sensor (Section 7.4) is used to obtain a value for the dissolved oxygen concentration (C_L). The main limitation of the oxygen balance method is the requirement for gaseous O_2 analyzers, such as a mass spectrometer, and accurate measurements of temperature and pressure.

∏ Use the following information to calculate the value of k_La in units of h^{-1}.

A culture was sparged with air at a rate of 0.5 l min^{-1} and analysis of input and exit gases by mass spectrometry showed that 300 mg l^{-1} oxygen was removed by the culture. The dissolved oxygen concentration of the culture, measured using a dissolved oxygen sensor, was 2.7 mg l^{-1} (partial pressure of oxygen in air = 0.209) (Henry's constant = 40 bar l g^{-1} oxygen).

Our approach is as follows:

Using Equation 7.5 [OTR = Q (Y_{in} - Y_{out})]

Y_{in} - Y_{out} = 300 mg l^{-1}

Q = 0.5 l min^{-1} = 30 l h^{-1}

So, OTR = 30 (300) = 9000 mg $l^{-1}h^{-1}$

Using Equation 7.3 [C^* = p_o/H]

C^* = 0.209/40 = 0.0052 g l^{-1} = 5.2 mg l^{-1}

Using Equation 7.4 [OTR = k_La (C^* - C_L)]

9000 = k_La (5.2 - 2.7)

k_La = 9000/2.5 = 3600 h^{-1}

For bench scale bioreactors, k_La values in the range 3000-4000 h^{-1} are common.

Sulphite oxidation method

An alternative method of k_La measurement is based on the oxidation of sodium sulphite to sulphate in the presence of a catalyst:

$$Na_2SO_3 + 0.5O_2 \xrightarrow{Cu^{2+} \text{ or } Co^{2+}} Na_2SO_4$$

The method involves filling the vessel with 1 mole l^{-1} Na_2SO_3, adding catalyst and commencing aeration. Samples of liquid are then taken at regular time intervals and the amount of unreacted sulphite is measured. Since the rate of reaction is very fast, the O_2

concentration in the solution equals zero and the O_2 transfer from the gas phase into the solution is the limiting factor. The value of k_La is then determined from:

Rate of sulphite oxidation = $k_La \quad . \quad C^*$

disadvantage
of sulphite
oxidation
method

Since a sulphite solution is used to simulate a fermentation broth the method gives no indication of the current oxygen demand by a culture. Large errors in k_La measurement can therefore be introduced. Consequently, the k_La value, determined by sulphite oxidation, is considered a characteristic of the bioreactor and the method is not suited for estimation of k_La values of individual fermentations. Another limitation of the sulphite oxidation method is the substantial cost of chemicals needed for k_La measurement of large bioreactors. However, the method has proved useful for comparing oxygen transfer characteristics in similar but different size vessels.

∏ Can you think of an advantage that the sulphite oxidation method has over the oxygen balance method?

The sulphite oxidation method does not require measurement of gaseous O_2 content. The method therefore does not require sophisticated and expensive equipment, such as a mass spectrometer or paramagnetic gas analyser.

Dynamic gassing out

This method is often used for the measurement of k_La values for actual fermentations. It involves stopping the air supply to a respiring culture and measuring the rate of fall of dissolved oxygen. The air supply is resumed before respiration is hindered by the fall in dissolved oxygen concentration. The increase in dissolved oxygen is then followed. Figure 7.2 illustrates the method.

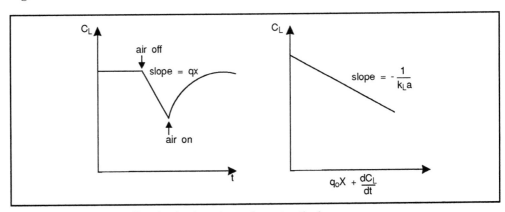

Figure 7.2 Measurement of k_La by the dynamic gassing out method.

The change in dissolved oxygen concentration is given by (you will not be expected to recall these equations):

$$\frac{dC_L}{dt} = k_La \, (C^* - C_L) - q_oX$$

which on rearranging becomes:

$$C_L = -1/k_La \, (q_oX + dC_L/dt) + C^*$$

This equation has the same form as the general equation for a straight line ($y = mx + c$; where m = slope and c = y-axis intercept). The value of k_La is therefore determined from the slope of a C_L against $(q_oX + dC_L/dt)$ plot (see Figure 7.2).

advantages and disadvantages of gassing out

The advantage of this method is that only one parameter (the dissolved oxygen concentration) has to be measured. However, a limitation of this method of k_La determination is that it assumes that the culture is rapidly degassed when the air supply is switched off; this is not always the case, particularly for mycelial cultures. Another disadvantage is that the results are inaccurate due to the long response time of the oxygen sensor and due to oxygen transfer from the head space of the fermenter. The latter source of error can be removed by exchanging head space air with nitrogen.

SAQ 7.4

For each of the following statements, select appropriate method(s) of k_La determination from the list provided below.

1) The k_La value is considered a characteristic of the bioreactor rather than of the fermentation process.

2) Requires measurement of gaseous O_2 content.

3) Requires multiple measurement of dissolved oxygen concentration.

4) Only one parameter has to be measured.

5) Oxygen saturating concentration (C^*) must be known.

6) Method influenced by response time of dissolved oxygen sensor.

The methods of k_La determination are: oxygen balance, sulphite oxidation, dynamic gassing out.

When the optimum process conditions are found at the laboratory scale there is a need to translate these findings for use in large bioreactors This is known as scale up and in aerobic systems it is most commonly performed on the basis of k_La values. Scale up at fixed k_La involves altering operating parameters such as bioreactor geometries, stirrer power and aeration rates. This approach ensures that the OTR does not decrease with increase in size of the bioreactor. k_La is dependent upon a variety of fermentation conditions, these include:

- volume and diameter of the bioreactor;

- stirrer power and type of aeration system;

- aeration rate;

- viscosity, density and composition of medium;

- the type of micro-organism;

- the temperature;

- the antifoam agent uses.

The precise mathematical relationships between k_La and these fermentation conditions are beyond the scope of this chapter. However, in the next section we will consider in general terms how dissolved oxygen can be controlled in aerobic fermentations. k_La is extensively covered in the BIOTOL texts 'Operational Modes of Bioreactors' and 'Bioreactor Design and Product Yield'.

7.2.3 Control of dissolved oxygen

Dissolved oxygen can be controlled by aeration, by mechanical agitation or by both in combination.

Mixing affects mass transfer of gas to liquid in three ways:

- it disperses the gas into small bubbles and thus increases the interfacial area;

- it increases the gas-liquid contact time by 'hold-up' of bubbles in the turbulent liquid;

- it decreases the thickness of the stationary liquid film.

All three effects will increase the value of k_La and hence the OTR (see Equation 7.4). In this section we will consider the different basic design of bioreactors, according to the way in which the fermenter contents are aerated and mixed. The main types are:

- stirred tank bioreactors, in which both mechanical stirring and aeration help to mix the culture;

- loop bioreactors, in which the culture circulates throughout the length of the vessel. This liquid circulation may be driven either by a propeller or by air; loop bioreactors can also be regarded as particular types of stirred tank or bubble aeration bioreactors.

Stirred tank bioreactors

The stirred vessel is the most widely used bioreactor. It is the best understood and most versatile. The agitation system consists of a motor-driven shaft that enters the vessel through a bearing housing and mechanical seal assembly. The shaft may enter from either the top or the bottom of the vessel. Some types of stirred tank bioreactors are shown in Figure 7.3.

Agitation in a stirred vessel can be either free or baffled. Baffles increase turbulence at the fermenter wall and prevents vortex formation (Figure 7.3).

flow patterns in stirred tank bioreactors

Virtually full turbulence can be obtained by means of four equally spaced baffles of width about 10% of the vessel diameter and running the full depth of the liquid. In order to aerate in the baffled system air must be introduced by a sparger below the liquid level. In the vortex system air is dragged into the liquid from the air space in the region of the impeller. There is intense turbulence in the region of the impeller with vortex agitation but outside the impeller region the culture is circulated with little turbulence.

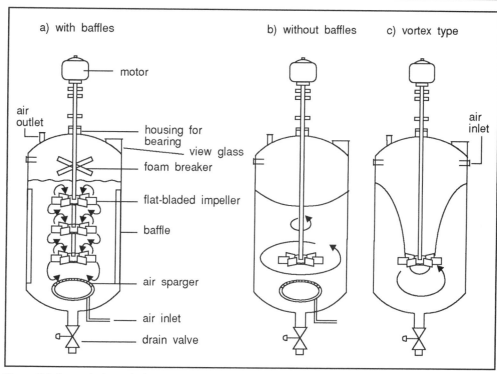

Figure 7.3 Some designs of stirred tank bioreactors. Baffles increase mixing, top-to-bottom of the vessel, by converting radial flow to axial flow, compare a) and b). Arrows show direction of liquid flow.

In the stirred bioreactor, the type and position of impeller used is also an important consideration. Figure 7.4 shows different types of impellers.

types of impellers

The disc stirrer is the most common type. This has 6-8 radial blades projecting out beyond the edge of the disc. The optimum diameter of the impeller is 25% of the vessel diameter. Vertical vanes cause radial flow of liquid in the absence of baffles and there is usually two or three such impellers (Figure 7.3a). In the case of two impellers the lower one is placed 0.5 impeller diameters from the bottom of the vessel and the second one 1.5 impeller diameters above the lower one. The spacing of impellers on the shaft is important to obtain the optimal flow pattern of liquid shown in Figure 7.3a. If the impellers are spaced too close together they behave like a single large impeller, which gives poorer mixing. If they are spaced too far apart stationary regions in the liquid may occur. In contrast to vertical vane impellers, inclined vanes cause an axial flow of liquid.

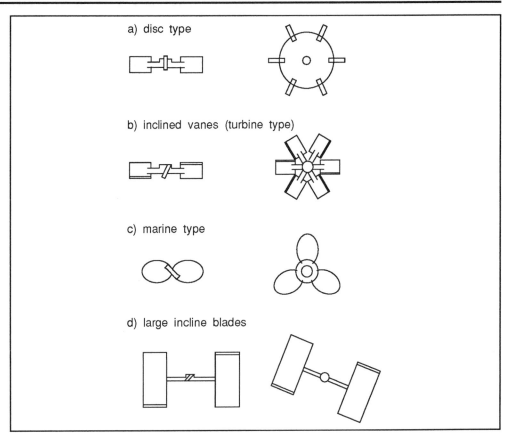

Figure 7.4 Types of impellers.

This last section has given you rather a lot of information about impellers and the way in which they mix liquids. It would be helpful if you check that you have understood this information, by doing the following intext activities. It will test your knowledge of the words used and the different flows that can be generated in a bioreactor. It will also test your understanding of the physical arrangement of impellers.

Π 1) Identify the component labelled 1) to 7) on the bioreactor. 2) Use arrows to show the direction of liquid flow during stirring. 3) Use the diameter of the vessel given in the diagram to determine the optimum lengths of A, B and C using the description given in the text.

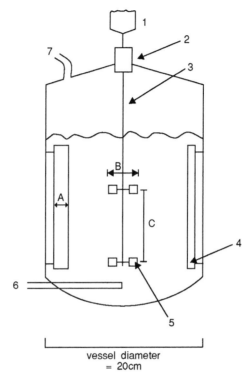

vessel diameter
= 20cm

The answer to 1 is that the component 1) drive motor, 2) housing for bearing, 3) drive (agitator) shaft, 4) baffle, 5) flat-blade impeller, 6) gas inlet and 7) gas outlet.

For the answer to 2) see direction of flow arrows shown in Figure 7.3a.

The answer to 3) is A = 20 x 0.1 = 2 cm (10% of vessel diameter), B = 20 x 0.25 = 5 cm (25% of vessel diameter), C = 5 x 1.5 = 7.5 cm (1.5 x diameter of impeller).

In the marine type of impeller (Figure 7.4), blades are curved and liquid flow is axial. Another type of impeller has large open inclined blades and compared to the disc and marine types requires less energy for equivalent oxygen transfer rates. Because large inclined blade impellers are not rotated as fast as disc type impellers, shear forces are reduced; this means that the mixing is gentler. Large inclined blades are therefore preferred for cells that are easily damaged during fermentation.

Π Round bottomed vessels are generally used with large inclined blade impellers. Can you think of a reason for this?

To prevent stagnation of liquid in corners at the low speeds of rotation.

Aeration in baffled, stirred tank bioreactors is by sparging. This is usually through a single nozzle or a perforated tube or ring positioned directly below the lower impeller.

∏ It is easy to see why oxygen transfer rate (OTR) increase with the rate at which air is supplied, but beyond a certain velocity OTRs are actually reduced. Can you provide an explanation for this?

At high air flow rates the gas bubbles coalesce to form 'slugs' (with small surface area/volume ratio) and so OTR is reduced. The actual rate beyond which this occurs depend on the degree of mixing in the bioreactor.

Bubble aeration bioreactors

In these types of bioreactors agitation of the culture is due solely to sparging with air at the base of the vessel; the movement of air bubbles to the top of the culture effects mixing. Agitation by bubble aeration is less efficient than mechanical agitation. However, the bubble aeration bioreactor has a number of notable merits, which are listed below.

1)   They are more economical to operate, since there is no requirement to drive impellers. This is particularly significant for large fermenters (eg 50 m^3), where the power requirement for mechanical stirring becomes a limiting factor.

2)   The absence of an agitator shaft and bearings reduces the risk of contamination.

3)   Agitation is much gentler than in stirred tanks.

As with stirred tanks the design of the vessel is important. Bubble aeration bioreactors have a greater height to width ratio, compared to stirred tanks. In order o prevent 'slugs' of air passing through the culture perforated baffles (sieve plates) are inserted at intervals throughout the length of the vessel (Figure 7.5). These break up the large bubbles into smaller bubbles and this increases OTR.

∏ Can you think of two reasons why OTR is greater for small air bubbles than for large air bubbles?

1) Small bubbles have a greater surface to volume ratio.

2) Small bubbles rise slower than large bubbles, so their residence time in the culture is greater.

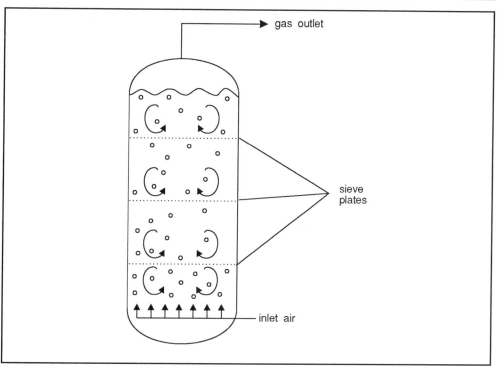

Figure 7.5 A bubble aeration bioreactor. Curved arrows show direction of liquid flow.

Loop bioreactors

The basic types of loop bioreactor designs are shown in Figure 7.6.

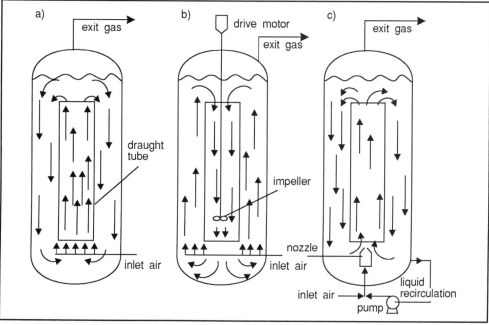

Figure 7.6 Basic types of loop bioreactors. a) Gas-lift type. b) Propeller type. c) Jet type. The arrows indicate direction of liquid flow.

The common feature of these designs is that culture is circulated continuously throughout the length of the vessel, with the upward and downward flows confined to separate sections.

gas-lift loop bioreactors

The simplest type, and the only one that is used commercially, is the gas-lift loop bioreactor. Since both aeration and agitation are effected by air entering the base of the bioreactor, the gas-lift type is considered a variant of the bubble aeration bioreactor. In the gas-lift vessel, however, sparging with air is confined to only part of the vessel - known as the draught tube. Movement of liquid up the draught tube occurs because it contains air bubbles and so has a lower density than the liquid in the non-sparged part. After gas disengagement at the top, the culture descends in the bubble-free section (downcomer). In some gas-lift loop bioreactors the downcomer is a separate tube external to the sparged vessel.

In the propeller loop and jet loop bioreactors, liquid circulation is assisted by mechanical means (Figure 7.6b and c). This improves OTRs but also has some disadvantages, such as increased shearing and high power consumption.

SAQ 7.5

For each of the following statements, select appropriate designs of bioreactors from the list provided.

Statements

1) The bioreactor usually has vertical baffles.

2) The bioreactor usually has only one impeller.

3) Gas (air) is sparged into the liquid culture.

4) The speed of impeller rotation is relatively slow.

5) Mixing is achieved solely by movement of bubbles.

6) The bioreactor does not have an agitator shaft.

7) Liquid flow is radial.

8) The vessel has a downcomer.

Design of bioreactor

Loop - gas-lift type.

Bubble aeration.

Loop - propeller type.

Stirred tank - vortex system

Stirred tank - open inclined blades.

Loop - jet type.

Stirred tank - flat blade impeller.

7.3 Temperature

temperature
sensing
In order to optimise growth, fermentations are usually carried out at constant temperature. Temperature sensing is not normally a problem as there is a good selection of sterilisable sensors commercially available. A thermistor or platinum resistance thermometer fixed inside the fermenter is usually used to generate a signal which is then used to regulate the temperature control system.

A constant temperature is only maintained if the rate of heat production is balanced by the rate of heat loss. This overall balance is given by:

Heat of agitation + heat of fermentation = heat of evaporation + heat of radiation + heat removed by cooling water.

Heat of agitation, in a stirred tank bioreactor, can be calculated from the energy consumption of the drive motor, minus the losses through the drive shaft and its assembly. The rate of microbial heat production (heat of fermentation) can be calculated by taking into account the growth rate and the reactor volume:

$$\text{Heat production rate} = \frac{\mu.X}{Y_{kjoules}}$$

(E - 7.6)

where:

μ = specific growth rate;

X = biomass concentration;

$Y_{kjoules}$ = heat yield coefficient

$Y_{kjoules}$ values are dependent upon the type of micro-organism and the type of substrate. Generally, less heat is produced for highly oxidised substrates compared with more reduced substrates.

Actual temperature control requires selection of an appropriate system according to the exact requirements of the fermentation. Figure 7.7 shows different types of temperature control systems.

cooling devices
Chilled or mains water generally provide adequate cooling via jackets, cooling coils, or indirect heat exchangers. In the case of internal coils their presence may cause the following problems:

- increased 'fouling', since the coils provide more surface for micro-organisms to adhere to;

- interference with the dispersion of gasses and possibly the oxygen transfer rate.

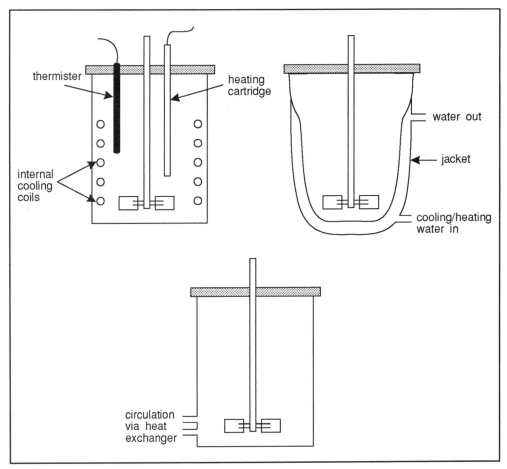

Figure 7.7 Systems for control of bioreactor temperature.

It follows that external cooling coils or jackets have the advantage that they do not clutter up the inside of the vessel. A limitation of external coils, however, is that the surface area for heat transfer (the surface area of the fermenter) may not be large enough to allow sufficient heat removal. This is explained by the heat transfer equation:

$$q = Ha\,\Delta T \qquad\qquad\qquad\qquad\qquad\qquad\qquad\qquad\qquad\qquad\qquad (E-7.7)$$

where:

q = heat transfer rate per unit volume of fermenter contents;

H = heat transfer coefficient;

a = surface area per unit volume;

ΔT = temperature difference.

∏   Can you think what heat transfer from fermentation broth to cooling surfaces is analogous to?

It is analogous to oxygen transfer from the gas phase to the cells in the fermenter (compare Equations 7.4 and 7.7).

As we saw for oxygen transfer, agitation also affects the efficiency of heat removal. Heat transfer rates at surfaces are controlled by the rate of transport of heat through the stagnant layer of liquid close to the surface. Agitation reduces the width of this layer and so reduces resistance to heat transfer. According to the heat transfer Equation 7.7, agitation increases the value of H. (Note that the physical principles of heat and mass transfer are discussed in the BIOTOL text 'Bioprocess Technology: Modelling and Transport Phenomena'.

∏ Explain why internal coils, rather than external coils, are more appropriate for large bioreactors.

The explanation is that, for large vessels, the surface area per unit volume (a) is smaller than that for small vessels. The heat transfer rate is therefore less and internal coils are used to increase surface area beyond that of the vessel surface. In the larger vessels there is also more room for the coils.

For heating, cartridge heaters can be inserted directly into the vessel but they sometimes suffer from medium 'burn-on' unless there is adequate mixing. Alternatively steam can be used which is an extremely rapid form of heating. However, if steam is passed directly into the jacket, internal coils or heat exchanger localised temperature gradients may be established.

7.4 Sensing the environment

Various sensors have been developed to monitor fermentation parameters. They can be categorised as those sensing 1) the physical environment, 2) the chemical composition of the environment and 3) the microbial cell concentration and its intracellular events. A list of measured parameters for fermentation processes and shown in Table 7.4.

Most of the biological parameters and many of the chemical parameters must be measured outside the bioreactor after sampling of liquid culture. However data on physical parameters and some of the chemical and biological parameters can be measured, without liquid sampling, using sensors which are part of the fermentation equipment ie on-line.

sterilisation of sensors

Since the requirement for sterility is usually strict in biotechnology, sensors should be steam-sterilisable to allow sterilisation of the bioreactor with the sensor in place (*in situ*).

Sensors which can be repeatedly steam-sterilisable are now available for pH, redox potential, dissolved oxygen tension and dissolved carbon dioxide tension.

Parameter	Method of measurement	Control action
Physical		
Temperature	Thermistor/Thermometer	Flow rate of coolant.
Pressure	Diaphragm gauges	Opening/closing outlet valves
Power consumption	Watt meter	Change agitator speed
Agitator speed	rev. counter	Change agitator speed
Gas flow rate	Flow meters	
Foaming	Conductivity	Addition of antifoam reagent
Chemical		
pH	pH electrode	Addition of acid/alkali
Dissolved oxygen concentration	Oxygen electrode	Change agitator speed or air flow rate.
Redox	Electrode	
CO_2/O_2 in exit gas	Mass spectrometry Paramagnetic gas analyser.	Change air flow rate
Substrate/product concentration	Liquid sample analysis or biosensors eg glucose sensor, ethanol sensor, formic acid sensor.	Addition of substrate (or adjustment of another parameter).
Biological		
Biomass concentration	Liquid sample or heat or gas analysis, turbidity, etc.	
Enzyme activities	Liquid sample analysis	
ATP	Liquid sample analysis or biosensor	
DNA and RNA	Liquid sample analysis	
NADH	Liquid sample analysis	

Table 7.4 Some typical parameters that can be measured in fermentation processes.

sampling without liquid transfer

Another method for assay of dissolved gasses and volatile medium components without liquid sampling is based upon immersion of a length of tubing, permeable to the compound of interest, in the culture. Continuous flow of carrier gas through the tubing sweeps the compounds which penetrate the tubing to a gas analyser. A limitation of this approach is the considerable delay in measurement (2-10 mins). The method is therefore not suitable for monitoring rapid or transient changes in concentration. The tubing method is illustrated in Figure 7.8.

biosensors

In recent years several biosensors have been developed for on-line measurement of organic components in the liquid phase. They consist of a biological recognition component (immobilised enzyme or cells) in close proximity or integrated with an electrode (signal transducer). A signal is generated in proportion to the extent of the reaction which, in turn, is proportional to the quantity of the substrate in the liquid culture.

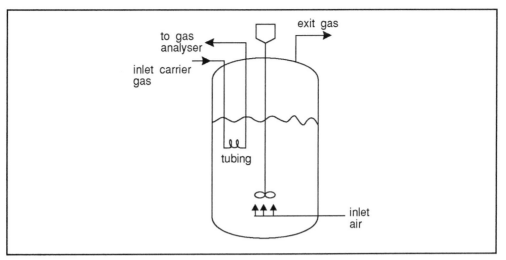

Figure 7.8 Measurement of volatile medium components and dissolved gasses by the tubing method.

The main problems with biosensors at present are their limited operating life and their failure to withstand steam sterilisation. In some cases however the latter problem can be overcome by pumping fresh biological material into the chamber behind the electrode membrane after sterilisation and cooling of the bioreactor (Figure 7.9).

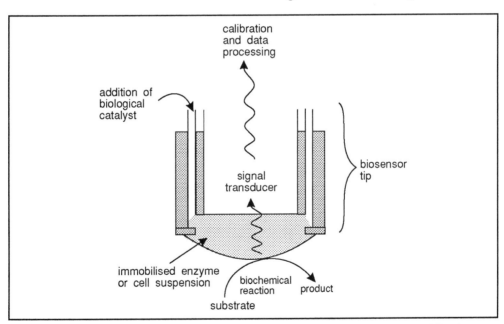

Figure 7.9 Activation of a biosensor following bioreactor sterilisation. The space for the immobilised enzyme or cells in the biosensor probe is left empty during sterilisation of the bioreactor.

∏ Name a method of on-line measurement for each of the fermentation parameters
    listed below (refer to Table 7.4 if you have difficulty with this).

Temperature
Dissolved oxygen
Exit CO_2 concentration
Gas flow rate
Glucose
Agitator speed
Biomass concentration
Pressure
Foaming

We will now consider in more detail the sensors used for the monitoring of pH and
dissolved oxygen concentration. These are now widely used for on-line data acquisition
in both laboratory scale and industrial scale fermentations.

7.4.1 Measurement of pH

The pH of fermentation processes is usually monitored continuously via a pH electrode
(probe). To avoid the use of an external reference electrode, the probe should preferably
be a combined glass electrode and reference electrode. Figure 7.10 shows a combination
pH electrode designed to be steam sterilisable.

Glass electrodes, such as the one shown in Figure 7.10, are made by fusing a membrane
of pH responsive glass across the end of a glass tube. The glass membrane behaves as a
hydrogen electrode when placed in aqueous solution and gives rise to potentials at the
glass liquid surface. The inner cell is usually a silver-silver chloride electrode in a
buffered chloride solution.

The type of pH electrode shown can be steam sterilised *in situ* only if the pressure inside
and outside the vessel remains equal during sterilisation. This, of course, would be the
case when the bioreactor vessel is transferred to an autoclave for sterilisation.

For *in situ* sterilisation by externally generated steam or by heating probes within the
vessel the pH electrode must include a housing to provide pressure balance during
sterilisation. In many cases this simply involves the use of pressure gauges and a
'bicycle pump' to manually equilibrate pH probe pressure with vessel pressure.

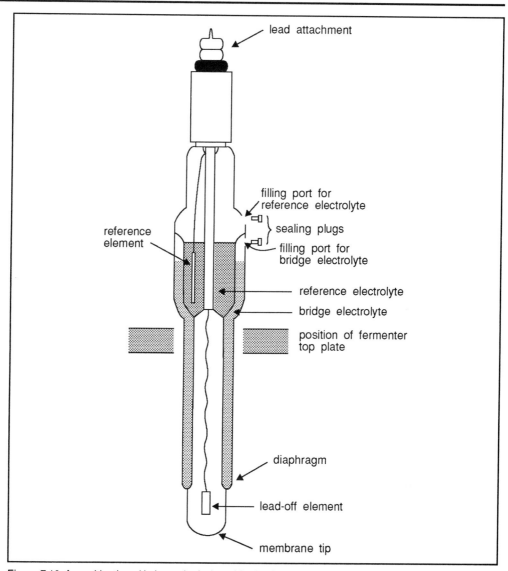

Figure 7.10 A combination pH electrode designed for *in situ* steam sterilisation.

7.4.2 Measurement of dissolved oxygen

oxygen probes

Although there are a variety of different designs of dissolved oxygen probes now available, there are essentially only two types: galvanic or polarographic types. These electrodes measure the partial pressure of the dissolved oxygen, ie the dissolved oxygen tension, not the dissolved oxygen concentration. In both types an oxygen permeable membrane, such as teflon, usually separates the electrolyte from the culture medium (Figure 7.11).

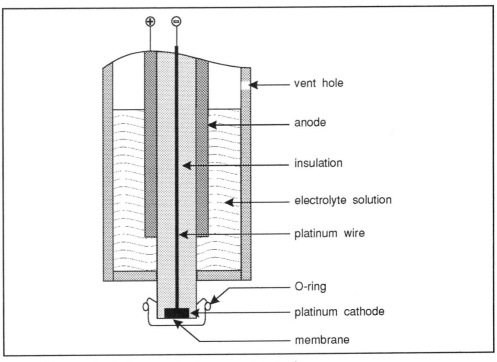

Figure 7.11 Stylised probe for measuring dissolved oxygen tension.

Both types also have a platinum cathode, where oxygen is reduced:

$$0.5\ O_2 + H_2O + 2e^- \rightarrow 2OH^-$$

In a galvanic type the anode is lead:

$$Pb \rightarrow Pb^{2+} + 2e^-$$

A small amount of current is drawn from the anode to provide a voltage measurement. This is correlated with the oxygen flux reaching the cathode surface.

In the polarographic type of oxygen electrode, the anode is silver and a constant voltage from an external source is applied:

$$Ag + Cl^- \rightarrow AgCl + e^-$$

In this type the resulting current which depends on the oxygen flux to the cathode is measured.

Both types suffer from 'drift' caused by:

• accumulation of hydroxyl or metal ions;

• depletion of chloride ions;

• external fowling of the membrane surface.

However, the polarographic type is generally favoured because of its greater reliability and robustness.

Another limitation of membrane-covered dissolved oxygen probes is that the response time is quite long at between 10-100 seconds. This is governed by a series of transport steps in which oxygen moves from the bulk liquid to the outer membrane surface, diffuses through the membrane, and finally diffuses through the electrolyte solution to the cathode surface. At the cathode, reaction effectively occurs immediately. The overall transport rate is usually determined by movement of oxygen to the outer surface of the membrane. The electrode output will therefore depend upon factors such as viscosity of the culture and hydrodynamic conditions near the tip of the probe.

SAQ 7.6

Identify each of the following statements as true or false. If false give a reason for your response.

1) A dissolved oxygen sensor is a direct measure of oxygen concentration.

2) The galvanic type of dissolved oxygen sensor has an externally applied voltage.

3) The polarographic type of dissolved oxygen sensor has a silver cathode.

4) Internal cooling coils in bioreactors increase the rate of heat transfer mainly be increasing turbulance.

5) Increased agitation increases the heat transfer coefficient.

6) Increased agitation may reduce the response time of a dissolved oxygen sensor.

7.5 Computer control in fermentation processes

There are a number of advantages to be gained by coupling fermentation process instruments to digital computers.

Π List some benefits of computer control of fermentation.

Now compare your list with the following major benefits:

• quick and efficient data management and storage;

• near exact reproducibility and maintenance of optimal conditions;

• detection of minute changes in a fermentation over very short intervals;

• improved optimisation of overall economics of the process;

• more fermentation plant flexibility in adapting to market demands.

In this section we will consider how computers can be used to effect control based on data generated by on-line sensors or by off-line measurements.

Two main types of computer control are now in use:

Direct digital control where the computer examines the input variable (signal) from the sensors and uses the information to produce a signal that is sent to the final control element (eg a control valve) to produce corrective action.

Digital set-point control where the computer adjusts the set-point but also produces a signal that is sent back to effect control.

Both types are forms of feed-back control, since control action is triggered by an error signal. By 'error signal' we mean the signal that is generated as a result of the measurement of a value by a sensor and the value that had been set. In digital control, each control loop consists of:

- the culture;
- measurement device (eg pH electrode);
- digital controller (computer);
- final control element or actuator (eg control valve or pump).

Figure 7.12 shows the elements of a digital feedback control system.

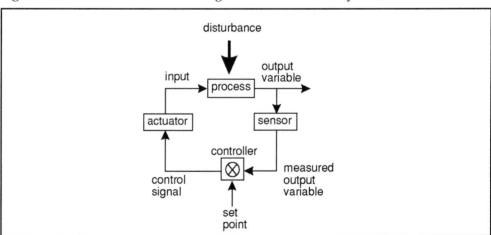

Figure 7.12 Elements of a feedback control system.

The corrective action is based on the difference between desired (set-point) and measured value and is governed by a control loop, of which there are several different types.

7.5.1 Types of control loops

The simplest type of control is on-off control. Here, the actuator is turned on when the error exceeds a specified value and is turned off when the error falls below another specified value, or vice versa. You will recall that this approach was described in Section 7.1 for control of pH by acid/alkali addition.

If the final control element provides a continuous range of outputs such as a variable-speed motor on the fermenter impeller shaft or a continuously adjustable valve on the air supply it is common to use proportional-integral-derivative (PID) control, or some variation of this.

A properly adjusted PID controller often provides excellent control of the measured variable. However, a poorly adjusted feedback controller can destabilise a fermentation by causing undesirable and accentuated fluctuations. Procedures for the tuning of control loops for best performance can be found in most textbooks on process control and will not be considered in this chapter. Aspects of process control and controllers are covered in greater depth in the BIOTOL text 'Bioreactor Design and Product Yield'. Figure 7.13 shows the behaviour of a typical feedback control system, using different types of control loops, when it is subjected to a disturbance.

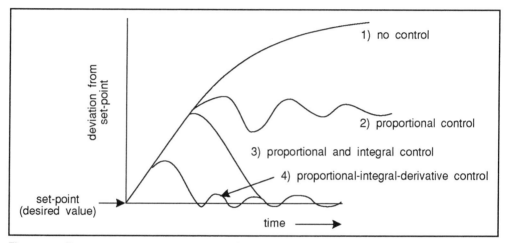

Figure 7.13 Response of a feedback control system.

With proportional control only, the control system is able to arrest the rise of the control variable and bring the variable back to a new steady-state level, which may not be the set-point value. The difference between the new steady-state value and the set-point value is called the offset. The addition of an integral control action eliminates the offset and brings the controlled variable back to the original value. This benefit of integral action is often balanced by a more oscillatory behaviour. Addition of derivative action to the proportional-integral action gives a definite improvement in the response:

• the rise of the controlled variable is arrested more quickly;

• it is returned to the set-point value more quickly;

• there is little or no oscillation.

∏   You might find it useful at this point to compare, in your own words, the different types of control loops mentioned so far.

So far in this chapter, we have considered what we might call 'univariable control', where variables such as pH and temperature are maintained at some desired value. What we have not considered, but which nevertheless is very important, is the interaction of process parameters. For example, growth rate and product yield are dependent upon a whole range of physical and chemical parameters, and the physical and chemical parameters themselves are also interrelated. For example, we have already seen that air flow rate will influence oxygen transfer rate, but it will also affect mixing and so both nutrient transfer and heat transfer (and thus temperature) will be affected.

Because many variables of fermentation processes are interrelated it is often more useful to consider controlling the fermentation so that some metabolic property, such as growth rate or product formation rate, is kept at a desired value. This type of control is

multivariable cascade control

termed multivariable cascade control and is becoming the predominant feature for fermentation monitoring and control. Here, the metabolic parameter is measured and compared to its set-point value and any error determines the 'metabolic control action'; this may involve control of one or more physical and/or chemical variables via feedback control loops of the type already described.

SAQ 7.7

Define, or explain the meaning of, the following terms in the context of fermentation process control.

1) Offset.

2) Set-point.

3) Error signal.

4) Actuator.

5) Feed-back control.

6) Direct digital control.

7) On-line sensor.

Summary and objectives

In this chapter we have seen that control of pH, dissolved oxygen concentration and temperature is necessary to maintain these fermentation parameters at their desired values. For automatic pH control the buffering capacity of the medium and strength of acid/alkali are important considerations. Temperature control is effected by external or internal apparatus; the choice of control system depends largely upon the size of the bioreactor and the extent of heating/cooling required. Provision of sufficient oxygen in aerobic fermentation is not a trivial task because microbial oxygen demand is high and oxygen solubility is low. High oxygen transfer rates can be achieved by increasing the value of k_La through aeration and agitation. Several designs of bioreactors can be used for this purpose. Factors influencing the choice of bioreactor include the oxygen demand and the susceptibility of the organism to shearing.

Reliable sensors, which can be steam sterilised, are now widely used for monitoring pH, dissolved oxygen and temperature in fermentations. Biosensors are available for some other fermentation parameters, but these are less reliable and often cannot be sterilised *in situ*. Signals from sensors can be evaluated by computers and used to effect digital feedback control of fermentations. Proportion-integral-derivative control is the most effective type of control loop.

Now that you have completed this chapter you should be able to:

• predict the direction of pH change for different types of cultivations;

• select appropriate methods of pH control and describe the interaction of factors influencing automatic pH control;

• describe Henry's law and factors influencing oxygen solubility;

• list merits and limitations of methods for determination of k_La;

• relate OTR to μ and to k_La;

• describe different designs of bioreactors and appreciate the relative merits/limitations of each (in relation to both aeration and temperature control);

• describe sensors for measurement of pH and dissolved oxygen;

• distinguish different types of computer control strategies.

Viruses

Viruses

Overview

Viruses are particles which are overall very different from all other biological entities. The study of viruses has to be relatively isolated because the rules which are applicable to basic cellular processes such as reproduction, self preservation and nutrition are totally inapplicable to these acellular structures. The chapter opens with a general introduction defining how and why viruses are different to all living material and follows with a brief history of this relatively modern branch of microbiology.

structure of the chapter

The section on the structure of viruses is relatively long reflecting their diversity: in contrast the section on viral classification, a topic about which we know relatively little, is very short. We then deal in depth with viral replication detailing the individual stages and showing the uniqueness of the processes involved. An in depth treatment of replication by lytic cycle of one particular virus, the T_4 bacteriophage follows and indicates the complexity and efficiency of the multiplication process. The following section offers a comparison to the lytic cycle of a bacteriophage demonstrating lysogeny - a more stable, lasting relationship between host bacterial cell and bacteriophage, probably maintained indefinitely. The emphasis here is predominantly on viruses which infect bacterial cells. Viruses which infect animal and plant cells are discussed elsewhere in the BIOTOL series (see 'In vitro Cultivation of Animal Cells').

The methods used for cultivation of viruses are relatively standardised, the main differences being due to the conditions required for the propagation of the host cells.

The sections on quantitation and purification of viruses are also relatively brief; it is often a matter of deciding in which order certain techniques are employed to yield the most accurate count or the highest purity.

A small section at the end gives a hint as to the tremendous potential of viruses as a servant of mankind when used as model systems for genetic engineering. The subject offers tremendous scope for innovation in the future.

8.1 Introduction

Viruses may be defined as a group of infectious agents or genetic elements. They consist of either DNA or RNA, never both, enclosed in a protein coat which may be surrounded by an additional outer layer. The presence or absence of the outer layer and its complexity vary enormously. The layer may contain carbohydrate, lipid or additional protein, some of which may be functional enzyme molecules. The overall structure is therefore acellular or non-cellular and viruses are totally incapable of reproducing outside a host cell. Reproduction is therefore dependent on host cells - these may be of bacterial, animal or plant origin - and the process always occurs within the host cell.

acellular nature of viruses

intra- and
extracellular
phases

Viruses must therefore be capable of existing within the host cell (the intracellular phase) and also for long periods of time outside the host cell (the extracellular phase). In the extracellular phase viruses are often referred to as virions or virus particles. The virus enters a host cell by one of a variety of means, the process being termed infection. Within the cell the virus occurs principally as a replicating nucleic acid unit which commands the host to synthesise the other components of the virion. In other words the viral structure fragments and the outer layer(s), and the protein coat become completely removed from the nucleic acid. Subsequently a variable but usually large number of virions are released.

∏ Given that the average bacterial cell, for example *Escherichia coli*, is around 2 micron (μm) long, could you make any guess as to the size of virus particles?

The obvious answer is that they are smaller than 2 μm but you could have gone further and indicated that they are very much smaller. Not only do they have to enter host cells but, as we mentioned earlier, they multiply within the host to produce a large number of new virions, the number depending on the virus infecting the cell.

Virions in fact range from around 0.02 μm to 0.3 μm in size. We have used μm or micrometre as the standard unit of length when discussing bacteria but when discussing viruses it is often advantageous to use the nanometre (nm or 10^9 metres). Thus the range 0.02μm to 0.30μm becomes 20 to 300 nm.

SAQ 8.1

Viruses may be said to be unlike all forms of living cells. Place the following statements into one of three categories: either a) viruses differ from all living cells because they, b) viruses differ from most living cells because they, c) viruses are similar to living cells because they,

Give reasons for assigning your statements to particular categories.

Finally, on completion of this task try to answer the question - are viruses living or dead? Give reasons for your decision.

1) cannot produce energy independently;

2) contain nucleic acid;

3) cannot reproduce independently;

4) contain either RNA or DNA but not both;

5) have a structure which dissociates during replication;

6) they are metabolically inert having neither biosynthetic nor respiratory function outside of the host;

7) always contain protein;

8) are somewhere between 20 and 300 nm in size.

8.2 The history of virology

The word viruses is of Latin origin and simply means venoms or poisons. It was used in the early and mid nineteenth century in a very general sense to indicate 'harmful agents' and at this stage included what we now know to be bacteria. Historically the incidence of human viral infections has been known for thousands of years though the origins of viruses are shrouded in mystery. Egyptian hieroglyphics dating back to before 2000 BC depict Egyptians deformed almost certainly by polio and the epidemics around AD 150 to 250 which severely weakened the Roman Empire were probably caused by measles and smallpox viruses. In more modern times smallpox caused the downfall of the Aztec Empire after its introduction by the Spaniards in AD 1500.

Cumberland filters

In 1884 Charles Cumberland developed the porcelain bacterial filter, an important event in virology because around this time it had been proved that many infectious diseases were caused by bacteria; agents which could be recovered by filtration. However several diseases were evident for which a bacterial cause could not be demonstrated. At the turn of the century, scientists working independently on the virus causing tobacco mosaic disease in plants and on the virus causing foot and mouth disease in animals showed that the infections were not of bacterial origin but were caused by agents so small that they could pass through a bacterial filter. They were not ordinary toxins, they were active at incredibly low dosages and could be transmitted from organism to organism in filtered material. Beijerinck termed the tobacco mosaic disease agent a 'contagium vivium fluidum' or 'living germ that is soluble'. Shortly after this, Loeffler and colleagues and suggested from their work on foot and mouth disease that the causal agent might belong to 'the smallest group of organisms as yet undetermined'.

bacteriophages and the work of d'Herelle

As other filterable agents were rapidly discovered the term 'filterable viruses' was used but over the years the term filterable has been dropped and the original wide ranging use of the word viruses has been confined to the agents which we have described in Section 8.1. Of particular relevance to us in this chapter was the discovery in 1915 of viruses which use bacterial cells as hosts. Two years later d'Herelle coined the term bacteriophage (literally bacteria eater) because, as we shall see later, they 'eat holes in bacteria lawns'. The word bacteriophage or its abbreviated version *phage* are routinely used but let us remember that they are alternative names for viruses.

Finally the chemical nature of viruses was partially elucidated in 1935 when Stanley managed to crystallise the tobacco mosaic virus and found that it contained only or mainly protein. Shortly afterwards the virus was separated into its protein and nucleic acid components and sometime later the importance of the relatively small amount of nucleic acid was established.

<table>
<tr><td>SAQ 8.2</td><td>Indicate whether the following statements are true or false giving the reason(s) for your decision.</td></tr>
</table>

SAQ 8.2

Indicate whether the following statements are true or false giving the reason(s) for your decision.

1) The invention of the Cumberland filter was a very important advance in the development of virology.

2) The terms 'filterable viruses' and 'viruses' are synonymous.

3) Viruses are probably of ancient origin.

4) Most of the early investigative work in virology was carried out using viruses which infect bacterial cells.

5) The early conclusions to be drawn from the work of Stanley in 1935 were particularly misleading.

6) Bacteriophages are viruses which prefer to use bacterial cells as hosts but may infect alternative plant cells.

8.3 The structure of virus particles

Before embarking on a study of virus cultivation, we have to learn something of the nomenclature and classification which themselves depend on the structure of viruses. As the whole subject of viruses is new to this text we shall now, at a fairly simple level, look at the structure of viruses.

Although virology is a relatively new science, principally due to the small size of viruses, they have the advantage that they are relatively simple, fairly stable and can exist in intact form outside of the host cell.

The primary objectives of the virion (a viral particle) are to find a suitable host cell, enter it and control the management of the host cell in order to produce new virions. The important component within the virion is obviously the nucleic acid and this must be adequately protected. In addition to providing protection, some mechanism to enable recognition of and entry into the host cell is required. Efficient packaging and organisation are necessary because, as indicated earlier, virions are very small ranging from around 20 to 300 nm in size. The smallest viruses are therefore little larger than ribosomes whereas the largest viruses are of the size of the smallest bacterial cells.

capsid
protein coat
nucleocapsid
capsomeres

The nucleic acid of all viruses is surrounded by a protective protein layer which is generally called a capsid though protein coat or protein sheath are sometimes used. The nucleic acid plus capsid is called the nucleocapsid and such structures are diverse and may be very complex. The capsid is made up of a number of individual protein molecules - protein subunits, protomeres or more commonly capsomeres. Capsomeres are arranged in a very precise, repetitive manner around the nucleic acid. There may be only one type of capsomere for a given virus but usually however there are several. As we shall see later though for reasons of genetic economy the general rule is 'the fewer types of capsomere the better'.

rods and
spheres

Where more than one structural type of capsomere is found, the different types associate in specific, repetitive fashion to produce morphological units. This regularity gives a symmetry to the virions such that rotation of the virion by a small amount displays a similar shape or form. There are two major shapes involved, rod shaped viruses and spherical shaped viruses.

8.3.1 Rod shaped viruses - helical symmetry

Rod shaped viruses display helical symmetry and exist basically as hollow, either rigid or flexible, protein cylinders in which the nucleic acid is found. One well studied example is the tobacco mosaic virus (TMV) which is shown in Figure 8.1. This is an RNA containing virus which contains 2,130 identical capsomeres, each of which contains 158 amino acids. The virion is a rigid tube 18 nm wide and 300 nm long. It seems that the capsomeres size and packaging determine the width of the virion. The length of the nucleic acid determines the length of the virion.

TMV

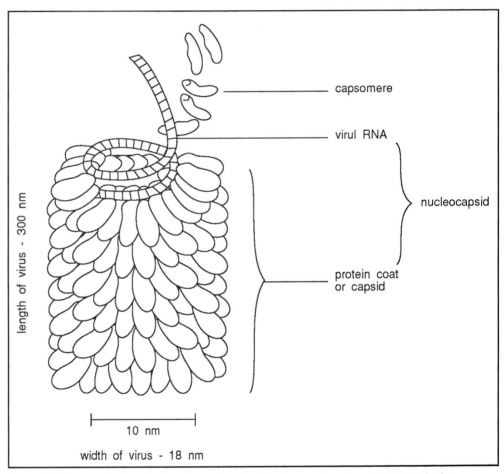

Figure 8.1 An example of a rod shaped virus showing helical symmetry - the tobacco mosaic virus.

∏   What is the advantage to the virion of requiring only a single or very few different types of capsomere?

The answer has solely to do with the genetic economy. There is relatively little space for the viral nucleic acid and thus, in practice, no facility to code for large numbers of different capsomere proteins and extra information as to how to construct the overall virion. In the tobacco mosaic virus for example only 474 of the 6000 available nucleotides on the RNA molecules are required to code for the capsomere proteins.

8.3.2 Spherical viruses - icosahedral symmetry

An icosahedron is a symmetrical structure having twenty faces and appearing virtually spherical by electron microscopy. These structures are very complex with each of the twenty faces showing a constant number of morphological units (remember, groups of structurally different capsomeres). The simplest icosahedral capsids have three morphological units per face whereas the larger, more complex ones can have up to twenty units per face. A simplified diagram of the shape is shown in Figure 8.2.

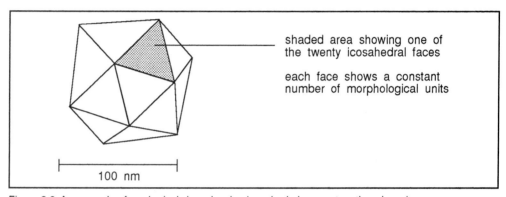

Figure 8.2 An example of a spherical virus showing icosahedral symmetry - the adenovirus.

8.3.3 Enveloped viruses

components of
viral envelopes

Many viruses, particularly animal viruses have envelopes surrounding the conventional helical or icosahedral nucleocapsid. These envelopes are membraneous, containing a lipid bilayer with glycoprotein. The glycoproteins are encoded by the virus but the lipids are derived from the host cell membrane. The envelope is clearly the part of the virion which will come into initial contact with the host cell and it is concerned with host specificity and initial penetration into the host. Remember that animal cells do not have a thick, rigid cell wall and virion uptake is often by endocytosis, thus if some of the envelope lipid is derived from previous host cells the potential host cell may be misled into thinking the virus is part of its own structure. An example of an enveloped virus is the herpes virus depicted in Figure 8.3.

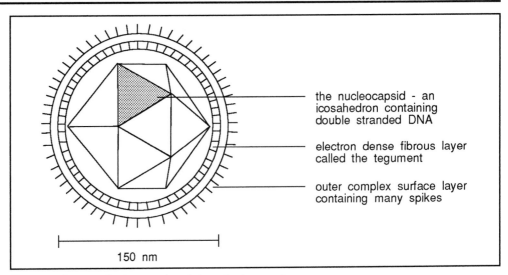

the nucleocapsid - an
icosahedron containing
double stranded DNA

electron dense fibrous layer
called the tegument

outer complex surface layer
containing many spikes

150 nm

Figure 8.3 An example of an enveloped virus - the herpes virus.

8.3.4 Complex viruses

Some virions are relatively complex; they may have complex capsids not displaying
helical or icosahedral symmetry and they generally contain one or more extra
structures. Many bacteriophages for example contain tails; we shall shortly examine the
life cycle of a bacteriophage which infects *E. coli* called the T_4 phage which contains an
icosahedral head with a helical tail joined to several appendages. A diagram of a T_4
phage is shown in Figure 8.4a. The tail is composed of a collar joining the head to a
helical sheath surrounding a central core. At the other end of the helical sheath is a
hexagonal base plate joined to tail fibres which are anchored by tail pins.

*T4 phage
structure*

Other complex viruses contain multilayered walls around the nucleic acid. For example
the vaccinia virus which is one of the largest viruses known (400 by 200 nm) contains a
nucleocapsid and two as yet undetermined structures in a very complex protein
envelope (Figure 8.4b).

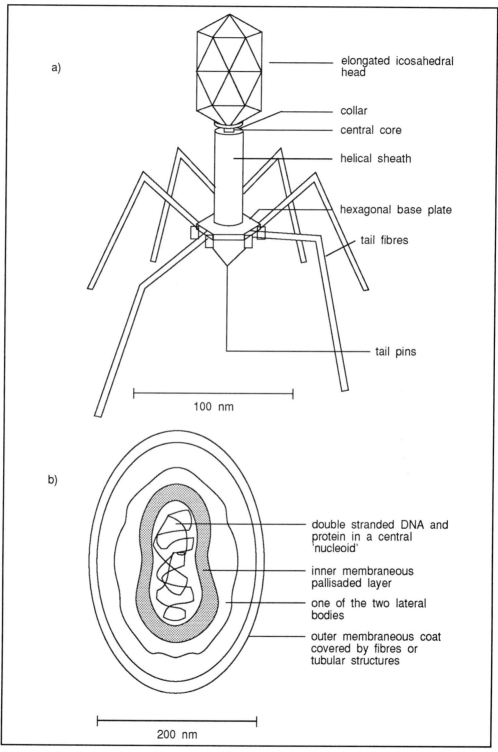

a)

elongated icosahedral head

collar

central core

helical sheath

hexagonal base plate

tail fibres

tail pins

100 nm

b)

double stranded DNA and protein in a central 'nucleoid'

inner membraneous pallisaded layer

one of the two lateral bodies

outer membraneous coat covered by fibres or tubular structures

200 nm

Figure 8.4 a) a diagram of a complex virus - the T$_4$ bacteriophage. b) Another example of a complex virus - vaccinia, a pox virus which infects animal cells.

8.3.5 Viral enzymes

It was originally perceived that all proteins within the capsid or envelope were non-enzymic. However some viruses, principally animal viruses and some bacterial viruses, do contain enzymes. It must be stressed that the virion is metabolically inert and enzymes are only activated on specific occasions.

∏ Can you think of perhaps two ways in which viral enzymes are used?

One role is to use this enzyme or enzymes to help gain entry into the host cell. For example some animal viruses have neuraminidase enzymes which help in breaking down the connective tissue of animal cells, and bacterial viruses possess lysozyme - an enzyme which degrades bacterial cell walls. The second role for viral enzymes is to aid in some way the process of replication of viral nucleic acid.

8.3.6 The viral genome

type and amount of genetic information in viruses

One outstanding feature of viruses is that they contain either DNA or RNA, never both. However within these confines a great deal of flexibility and diversity is evident. The genome may be double or single stranded DNA or double or single stranded RNA. The nucleic acid may be present as a single molecule or as several (upto ten or more) molecules. The relative content varies from up to 50% in non-enveloped viruses to 1 to 2% in enveloped types. In addition, the total amount of nucleic acid varies from around 10^6 daltons (to code for three to four proteins) through to 1.5×10^8 daltons (to code for over a hundred proteins), the latter in complex viruses such as vaccinia, herpes and T_4 bacteriophage.

As a generalisation, plant viruses usually have a single stranded RNA genome; animal viruses have any one of the four possibilities and whilst bacteriophages may be any one of the four, they are usually double stranded DNA types.

Much is now known of the nucleic acid structures of viruses and the process of virus multiplication - particularly from research into bacterial viruses. In addition the use of viruses as convenient tools for genetic engineering is well established. Although we shall mention this topic very briefly at the end of this chapter, our primary purpose is to study viral replication and cultivation.

SAQ 8.3	We have used a large number of names in Section 8.3 most of which are probably new to you. Before continuing to the next section, attempt to complete the following sentences. Try the exercise without looking at the prompt words at the end of the question, only referring to the list if you really get stuck!

1) The virus particle which exists outside of the host cell is called a [   ].

2) The four major structural groups of viruses are: [   ] shaped showing [   ] symmetry; [   ] shaped showing [   ] symmetry; [   ] viruses and [   ] viruses.

3) The viral nucleic acid is usually surrounded by a protein coat also known as a protein [   ] or a [   ].

4) The individual protein subunits in the protein coat are called [   ].

5) Virus particles which infect bacterial cells are called [   ] or [   ] for short.

6) An icosahedral structure exhibits [   ] symmetrical faces.

7) A repeating structure containing two or more different capsomeres is called a [   ].

Prompt list: bacteriophages, capsid, capsomeres, morphological unit; complex, enveloped, helical, icosahedral, phages, rod, sheath, spherical, structural unit, twenty, virion.

Finally, can you give an example of each of the four major structural types of virus?

8.4 Viral classification

Viral classification in the taxonomic sense is in its infancy compared to the classification of either prokaryotic or eukaryotic systems. Viruses are, for convenience separated into large groups based on one or two simple parameters and the unsatisfactory state of affairs exists where we know a lot about some viruses but relatively little about most of them.

SAQ 8.4	Consider the properties of viruses which we have studied so far. Can you firstly recall any which you think might be of importance in identifying viruses and secondly put them in possible order of importance?

8.5 The stages involved in viral replication

In this section we shall look very generally at the stages involved in the replication of animal, plant and bacterial viruses. Section 8.6 will then consider the replication of a named bacteriophage in greater detail.

Consider a free virus particle or virion recently released from a host cell.

<table><tr><td>

SAQ 8.5

</td><td>

List, in order, the stages in the process of viral replication. You should be able to identify at least five or six essential stages. After you have done this, make certain you read our response at the end of the text.

</td></tr></table>

We shall now consider each of these stages or phases in a little more detail.

8.5.1 Location of a suitable host cell

passive process

Location of a suitable host cell is a passive process in that viruses do not have the means to actively seek out potential hosts. Contact is made by chance during random movement through aqueous solutions or through the air. Thus the more virions and host cells there are, the more likely the chance of contact and successful infection.

8.5.2 Attachment

We have noted earlier that certain virions, particularly bacteriophages exhibit a high sometimes absolute specificity for a particular host. The specificity usually resides in this process of attachment of virion to host cell.

receptors and receptor sites

The virion has one or more proteins on its outer surface which react with specific surface components of the cell called receptors or receptor sites. These sites are normal surface components of the cell and vary according to the nature of the virus. For example cell wall lipopolysaccharides, proteins and the pili or flagella of bacterial cells can act as receptors for different bacteriophages.

∏ Can you decide what would be the virion attachment point of the T4 phage shown in Figure 8.4a?

You may have thought the head was the right answer but the correct answer is the base of a tail fibre which makes contact with an appropriate receptor site. Then the other tail fibre bases make contact establishing firm attachment.

If the receptor site is altered even slightly due to genetic change or removed altogether the host cell becomes resistant to infection by the virus as the virion cannot attach. The story continues however as mutant viruses are known which can overcome host receptor site modifications and so can again infect the host cells.

attachment of animal viruses

The absorption of animal viruses is similar to that of bacterial viruses and depends on the presence of receptor sites. The knowledge of the distribution of receptor sites has helped our understanding of the nature of viral infections of higher animals. For example polio virus receptors are only found in the nasopharynx, gut and parts of the spinal cord of humans whereas measles virus receptors are found in cells of most human tissues. The attachment part of the virus may be a capsomere structure or special structure such as the spikes of some enveloped viruses (see Figure 8.3).

One major difference between bacteriophage and some animal virus attachment processes is that animal viruses may be taken in by some type of accidental or intentional endocytosis (engulfment mechanism). Thus there may be some lack of specificity.

attachment of plant viruses

The situation with plant virus attachment and entry is far more varied. Some endocytotic mechanisms occur as well as entry via mechanical damage to cellular structures.

8.5.3 Penetration

The mechanism of penetration of viruses or parts of viruses into the cell is constant for a given virion but varies widely when looking at viruses as a whole.

∏ Can you write down any factors which are important contributors in determining the type of penetration used?

The two main factors are firstly the nature of the host cell outer layer - for example whether or not a thick cell wall is present. Secondly and more important the nature of the virion itself - is it simply a nucleocapsid or is it enveloped, complex etc.

bacteriophage into host by an injection mechanism

We shall shortly be looking at an example of a complex bacteriophage penetration which occurs by a complicated injection mechanism. Bacteriophages virtually always have to cope with complex bacterial cell walls and penetration is usually via a tiny orifice injecting only the nucleic acid. In contrast the penetration by endocytosis of animal or plant viruses often involves the whole of the virion. The process may take from minutes to several hours compared to the relatively speedy injection of phage material.

enzymes may aid penetration

Occasionally enzymes are found in the virion and during attachment these may be activated, their purpose being to aid entry into the host cells. For example bacteriophages may contain lysozyme and animal viruses may contain neuraminidase involved respectively in cell wall and cell membrane degradation.

We should also remember that if the whole virion penetrates into the cytoplasm an extra step is required, that of uncoating or removal of the outer layers of the virion to release the viral nucleic acid.

8.5.4 Replication

A substantial amount of knowledge has been gained regarding replication of the nucleic acid material. This knowledge has arisen from the realisation that virus replication can be used as a model system to study not only 'viral genetics' but also as a powerful tool for genetic engineering. We shall mention genetic engineering in Section 8.11 - the potential rewards to be gained by tampering with cellular genomes are immense.

A complicated study of the genetic mechanisms of viral replication is beyond the scope of this chapter however we shall examine the overall principles involved.

The subject is complicated by the possible incidence of several nucleic acid types within different viruses, namely - double or single stranded DNA or double or single stranded RNA.

+ and - forms of RNA

RNA can exist in one of two forms in the sense of firstly that which is required as a template for DNA synthesis (known as the + form) or the opposite form, that is the chemical 'mirror image' of it (known as the - form). Whatever the nucleic acid form in the virion, the + form of messenger RNA is required to initiate production of viral nucleic acids and proteins.

The following mechanisms seem to occur to produce the required + messenger RNA molecule. Remember that the process of transcription means the formation of messenger RNA which contains a complementary copy of the information stored in the DNA molecule.

Viral nucleic acid	Route taken	Required nucleic acid
± double stranded DNA	via transcription of - DNA strand	+ mRNA
+ single stranded DNA	via ± double stranded DNA; - DNA strand then transcribed	+ mRNA
+ single stranded RNA	acts as it is	+ mRNA
- single stranded RNA	+ RNA made directly	+ mRNA
± double stranded RNA	- RNA strand transcribed	+ mRNA

In addition one group of animal viruses have + single stranded RNA but produce + single stranded RNA in three stages via DNA. (+ RNA → - DNA single strand → ± double strand DNA → mRNA).

The synthesis of the + mRNA can begin very quickly, for instance + mRNA is detectable within minutes after the injection of DNA by the T$_4$ bacteriophage.

Once the mRNA has been produced then synthesis of the various components for producing complete virions begins quickly. The proteins synthesised are often grouped into one of three categories depending on when they are produced.

early proteins The early proteins (early enzymes) are almost always enzymes which are only produced in small quantities and are required either to aid viral nucleic acid synthesis or to inhibit some normal host metabolic process. Sometimes these are split into early and intermediate proteins.

late proteins
lytic proteins The late proteins are those synthesised in much larger amounts being the structural components of the new virions. Finally there may be lytic proteins produced to help lysis of the host cell and or membrane.

Π It may be helpful for you to draw on a sheet of paper a flow scheme to illustrate the stages of viral replication from penetration onwards. We have begun such a scheme below:

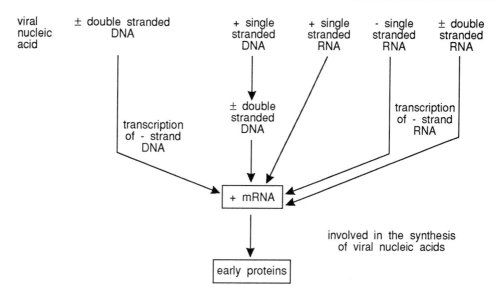

8.5.5 Assembly

A great diversity of methods of assembly are possible and to a large extent the complexity of the assembly process will depend on the complexity of the virion. In many instances capsomeres spontaneously assemble to produce capsids followed by nucleic acid insertion to form a nucleocapsid. Enveloped or complex viruses require extra terminal stages, the pox virus for example begins by forming a 'host membrane vesicle' and components then pass into this structure which eventually becomes the virion. Complex bacteriophage assembly is very complicated; the head, tail, and tail fibre assemblies occur independently of each other. Eventually the mature head containing nucleic acid joins to a preformed tail consisting of helical sheath, core and tail plate. Finally the tail fibres are added.

8.5.6 Release of virions

Most bacteriophages lyse their host cells generally following the production of enzymes to attack the cell membrane and/or cell wall. The whole process from attachment to lysis by a T_4 phage of *E. coli* can occur within twenty five minutes releasing up to 300 new T_4 virions.

Animal virions can be released similarly but to a much slower time scale (usually by non-enveloped viruses). Other phages and non-enveloped animal viruses may be released without cell lysis, the host cell continuing to produce virions and to grow, albeit at a reduced rate.

8.6 Replication of the T_4 bacteriophage in *E. coli* - the lytic cycle

This section is intended to describe in detail one particular phage multiplication sequence and to act as a revision of the last section. The term lytic cycle is used because the T_4 phage progeny are always released following host cell lysis.

| SAQ 8.6 | Try to draw from memory an 'exploded' diagram of a T_4 phage noting in particular all of the various structures which are required to produce a new virion. |

∏ At this stage it would be useful for you to repeat SAQ 8.5 as a revisionary exercise before continuing.

requirements
for attachment

Having established the individual stages we can now follow the sequence of events involving the *E. coli* T_4 phage. As indicated earlier, one or more of the bases of the tail fibres comes into contact with the receptor site(s) on the host cell. Very quickly all tail fibres become anchored and the base plate settles onto the cell surface. Note that with some phages, attachment requires the presence of particular ions (especially Ca^{2+} and Mg^{2+}). Others require particular biochemicals and amino acids. Clearly these components are involved with the molecular interaction between the phage and the receptor sites.

The base plate and helical sheath then undergo conformational changes so that the sheath contracts. This has the effect of forcing the core through the bacterial cell wall. ATP is known to be present within the T_4 phage and this probably drives the contraction process. In addition a tiny amount of lysozyme is found at the tip of the core and this becomes activated making a tiny hole in and through the wall. The DNA is then injected into the cell. There may be many virions attached to a single cell at the same time and many may inject their DNA almost synchronously. The process described so far and shown in Figure 8.5 occurs in minutes.

Once inside the *E. coli*, phage nucleic acid (double stranded DNA) initiates production of + mRNA (from the - DNA strand) within 2 to 3 minutes. Almost immediately production of host DNA, RNA and protein ceases. Within five minutes viral DNA synthesis begins. The scene is now set for controlled production of all of the components of the T_4 phage in the correct order and in the correct stoichiometry.

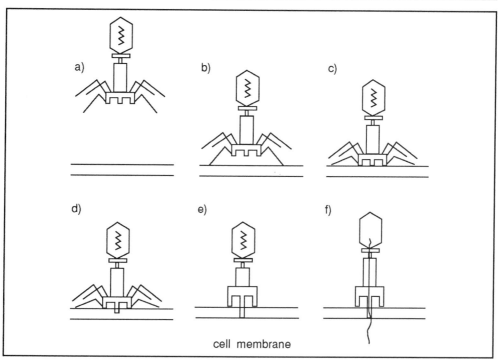

Figure 8.5 The location, attachment and penetration of *E. coli* cell by T₄ phage. a) virion near but not in contact with host cell membrane. b) virion landed on surface. c) virion attaching by base plate to cell membrane. d) tail contracting. e) lysozyme produced, tail fully contracted and penetration begins. f) DNA now injected into the host cell; the rest of the virion remains outside of the cell.

Assembly begins as shown in Figure 8.6 and within a given cell up to 300 separate phages will be produced. The insertion of the DNA is worthy of comment in that the DNA length is about 500 μm and the length of the head is approximately 0.1 μm - efficient packaging indeed! The construction of the first virion is completed after about fifteen minutes and, following production of lysozyme to digest the cell wall, lysis occurs after about 22 to 30 minutes. Overall a highly organised, efficient process in which a virion multiples up to 300 fold in less than half an hour.

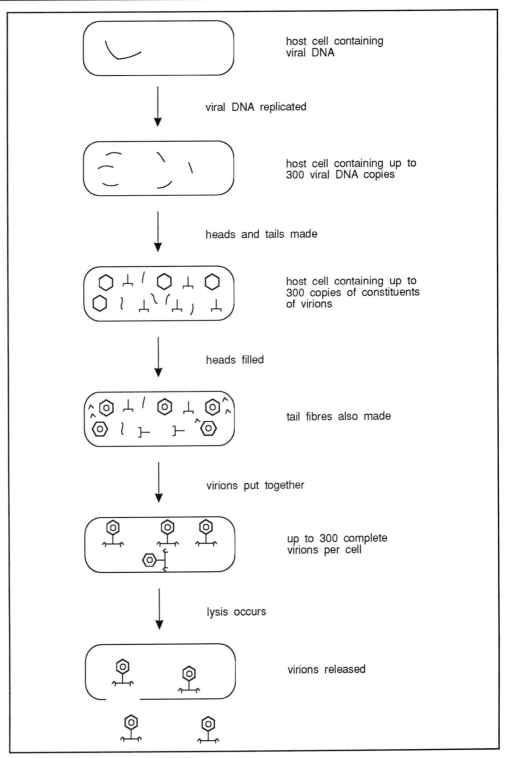

Figure 8.6 Replication, assembly and release of T₄ phage from *E. coli*.

8.7 The phenomenon of lysogeny

avirulent and
virulent phage

temperate
phage

lysogeny

prophage

We have so far confined our discussion almost entirely to a consideration of bacteriophages which gain release by lysis of the host cell. Such phages are sometimes called virulent bacteriophages. Some bacteriophages are capable of forming a relatively stable relationship with the host cell. Such bacteriophages are said to be avirulent or temperate bacteriophages. In these cases the phage first attaches and penetrates the host cell. The phage nucleic acid is then usually incorporated into the host genomes and is replicated with it. In some cases however the viral nucleic acid remains within the cytoplasm. This form of phage: host relationship is called lysogeny and is illustrated in Figure 8.7. The bacteria are said to be lysogenic and the phages are called temperate phages. The phage in the host cell in this latent or dominant form is called a prophage.

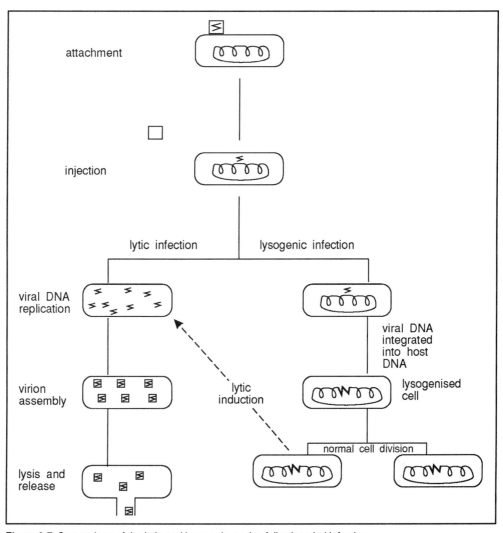

Figure 8.7 Comparison of the lytic and lysogenic modes following viral infection.

induction

The phage in the prophage state may be stimulated to reproduce (a process called induction). In this process, the prophage is converted into a virulent bacteriophage and enters the lytic cycle.

Once lysogeny has been initiated then that bacterial host cell is immune to infection by further phage of the same type - this is advantageous to both sides as we shall see shortly.

In fact most bacteriophages apparently can exist or even prefer to exist as temperate phages, a situation which confers advantages. In a culture containing temperate phages there are generally a small proportion (one in a thousand to one in a million) of virulent phages.

∏ Can you give a reason why lysogeny would be advantageous, for example in times of adverse environmental conditions?

In times of nutrient shortages bacteria will use up all of the available nutrient material and then degrade their own non-essential structures to provide energy for basal metabolism. Molecules such as mRNA would be included as non essential materials at this time. A lytic phage would then be in serious trouble - the host cell is no longer capable of producing new virion parts and in fact, those already produced within the cell are in danger of being degraded.

∏ Can you give the reason why immunity from infection by other temperate phages is advantageous - particularly in situations where large numbers of virus particles compared to bacterial cells are present?

In situations where there are large numbers of viruses then the likelihood of all host cells becoming infected and killed is high. Thus very quickly a situation would arise with lots of free virions but no host cells being available. If however a lysogenic relationship is set up barring entry of phages of the same type then a permanent relationship where the phage is protected is much more likely. A further advantage of course is to give protection against the small percentage of lytic phages within any given culture.

In situations where viruses can only be lytic then the virus/host relationship must be somewhat temporary due to the capacity of the virions to invade, multiply and lyse the host cell with such speed and efficiency.

∏ If for some reason the host cell DNA is damaged, for example by ultraviolet irradiation what happens to the lysogenic relationship?

induction in λ phage

The answer is that in some cases the prophage will become induced the process being called induction. For example the lambda (λ) phage of *E. coli* leaves the bacterial genome and becomes lytic, encouraging production of new lambda phage. Remember that although the cell may be irreversibly damaged, metabolism may proceed for several hours during which phage reproduction can occur.

Re-examine Figure 8.7 which shows a summary of the events which occur in a lytic or lysogenic process. The left hand side sequence shows a typical lytic process as discussed in Sections 8.5 and 8.6. The right hand side shows a lysogenic cycle in which the viral nucleic acid is incorporated into the bacterial genome. The connecting link or lytic induction shows the effects of damage, for example ultraviolet irradiation damage, on this particular lysogenic relationship. As mentioned certain temperate phage types, for example P1 of *E. coli* exist as a prophage separate and distinct from the host DNA; in other words as a type of plasmid. It is closely related however and remains at a ratio of one prophage per cell through many normal division cycles.

P1 phage

In the last three sections we have covered the replication of viruses using either a lytic cycle or lysogeny. The main emphasis was to concentrate on bacteriophage mechanisms and the lytic cycle of the T_4 phage of *E. coli* was studied in some depth. Many of the terms used were probably new to you, particularly those in Section 8.7 dealing with lysogeny. Thus the SAQ following will be a particularly useful exercise at this point to revise the nomenclature used in virology before we move on to Section 8.8, the cultivation of viruses.

SAQ 8.7

Complete the following sentences if possible avoiding looking back through the text.

1) The specific surface components of host cells with which virions react are called [   ].

2) The enzyme lysozyme is present in certain phages and its function is to [   ].

3) The term 'uncoating' means [   ].

4) From attachment of virion to host through to the lysis of the host can take as little as [   ] minutes and can produce up to [   ] new T_4 phages.

5) A bacteriophage which causes release by lysis of the host cell is sometimes called a [   ] bacteriophage.

6) Phages participating in lysogeny are called [   ] phages and the latent or dormant form of the phage is often called the [   ].

7) Lysogeny would be more advantageous than the lytic cycle for a virus because [   ].

8) The process of [   ] occurs when lysogeny converts to a lytic cycle.

8.8 The cultivation of viruses

As we have noted earlier the cultivation of viruses poses special problems because the viruses are unable to reproduce independently of living cells. Thus we have to provide not only a suitable physical environment but also host cells to which the viruses can attach, penetrate and bring about multiplication.

The conventional way of cultivating animal viruses was, for many years, to use fertilised hens' eggs which had been incubated for a few days after laying. A tiny hole was made through the shell using a sterile drill, the growing embryo was then inoculated and the hole sealed. By no means all viruses would grow in the embryo of hens eggs at all and several that did were site specific within the egg. The disadvantages of using this method are essentially practical in that there is a mixture of host cells types, the eggs are relatively fragile to handle, contamination may be a problem and finally virus purification following multiplication is often tedious.

Nowadays a tissue or cell culture technique is generally used in which a suitable animal cell type is cultured as a monolayer in a special container. Highly developed growth media coupled with the addition of antibacterial and antifungal agents collectively help to promote growth of pure cultures of the required animal cells. Inoculation by viruses followed by overlaying with agar to avoid spread of viruses results in virus multiplication in localised areas. Where virus multiplication occurs this may be accompanied by cell lysis, visible as clear patches or plaques. In some instances there is no lysis but changes to the cell structure called cytopathic effects may occur.

Plant viruses may be cultured in a variety of ways using whole plants, cell culture techniques or protoplasts. One method of inoculation of whole plants is to rub the leaf with a mixture of virus suspension and an abrasive such as carborundum. Viruses rely on the abrasive to partially break the thick carbohydrate cell wall of the leaf cells exposing the plasma membrane allowing host cell infection. A variety of localised effects may occur from necrotic lesions due to cell death to lesser changes in pigmentation or leaf shape.

Tissue culture of plant cells obeys the same principles as that for animal cell work. As we have mentioned earlier, plant cells, unlike animal cells, have a thick cell wall. This can be removed to produce a 'cell-wall-less' cell called a protoplast which allows easier virus penetration.

The culturing of bacteriophages is much easier than the culture of animal or plant viruses because the cultivation of the host cells is so much easier. Large, virtually unlimited quantities of bacterial cells may be obtained which can be infected with virions which undergo rapid multiplication causing almost total lysis of host cells very quickly. Growth can also occur on solid medium in similar fashion to that described for animal cells.

8.8.1 The one step growth experiment

The so called one step growth experiment is a convenient way of producing large quantities of viruses in a synchronised manner. The rationale behind the experiment is quite simple and the results are quite dramatic.

If, for example, a culture of the T_4 phage of *E. coli* is required, a suitable culture of young, actively growing *E. coli* is mixed with virions and left for a few minutes so that virions can attach to the host cells. The culture is then diluted many fold simply to avoid or lessen the possibility of any newly produced viruses from infecting further cells.

Figure 8.8 shows the relative virus count (in plaque forming units or PFUs) against time. Various stages or periods are named. A PFU is a measure of the number of intact viruses - we will give more explanation later)

SAQ 8.8

From your knowledge of the sequence of events of the T₄ lytic cycle (Section 8.6) try to explain briefly what exactly is happening in each period.

Try to answer the question without looking at the following hints but if you get stuck, take them one at a time.

Hints:

1) The processes indicated in the boxes refer to the production within the host cell of the building blocks for new viruses.

2) The dramatic increase in relative plaque count should indicate to you what is happening to the host cells.

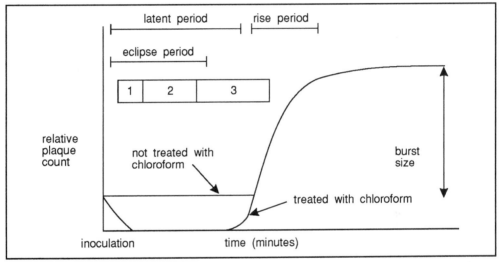

Figure 8.8 The stages involved in the one step procedure for virus cultivation. 1 early enzyme production phase; 2 viral nucleic acid production phase; 3 production of the remainder of the virion structures phase. Note that burst size is the ratio of the number of viruses produced to the number of viruses in the innoculum. Thus a burst size of 100 would indicate that 100 viruses were produced from each virus which successffully involved a host cell (see SAQ 8.8 response and the text for further details).

Let us now go through the experiment adding a little more detail to the answer to SAQ 8.8.

Remember that the whole time scale for a T₄ phage lytic cycle occurs in well under 30 minutes in optimum conditions so the events occurring are rapid and largely in synchrony. At time zero the viruses are added to a suitable exponentially growing *E. coli* culture. After a few minutes the culture is diluted significantly but already the viral nucleic acid will have entered the host cell and initiated + mRNA and early enzyme eclipse phase production. During this early part of the latent period, called the eclipse phase, there are no mature virions within the host cells and no capacity for 'viral infectivity'. Thus if we kill the bacterial hosts with chloroform we will find no intact, viruses (see Figure 8.8). Viral nucleic acid is present within the host cells but this would not survive in its naked form if for any reason the host cell was damaged at this point. As the eclipse period ends, the host cell has now begun its final phase - that of production of the viral components - and very quickly new virions are constructed. For a short time until the end of the latent period there is no cell lysis and thus no increase in the plaque forming units. Finally the rise period occurs which is relatively short as the cells lyse almost in synchrony and this period is characterised by a significant increase in plaque forming units.

∏ Why was the culture diluted after initial mixing of host cells and virions?

Inevitably there would be some uninfected host cells remaining after the initial infection process. If the culture had not been diluted, the likelihood of the new virions infecting such cells would be high. This would be undesirable as the object is to produce new virions at the same time, not to remove all host cells. Dilution largely overcomes this problem by significantly lowering the chance of virion to cell contact.

length of viral replication cycle

The whole cycle for bacteriophage is generally completed within 30 minutes to one hour but for animal cell viruses it is much slower, usually from twelve to twenty four hours though up to fifty hours is common. The number of new virions however which can be produced from animal cells can approach 1000 per cell, much more than are obtained per bacterial host cell.

8.9 Measurement of viral numbers

As an essential ingredient to furthering our knowledge and understanding of viral replication and mode of existence we must be able to satisfactorily estimate viral numbers.

use of electron microscope to estimate viral numbers

There are essentially two methods, either direct counting using microscopy or measuring the infectious unit concentration. Virtually all viruses are too small to be seen by the light microscope thus an electron microscope has to be used. The electron microscope has the disadvantage of being unsuitable for day to day routine techniques. Where it is deemed essential to carry out a direct particle count however, for example to construct a calibration curve, one can mix the virus particles with conveniently sized latex beads. If the relative numbers are similar to each other when viewed the ratio of viruses: beads can be found and, if the concentration of latex beads is known, the concentration of virions can be calculated.

It is usual however to measure viral numbers in some way by monitoring their effect on host cells. It is common to quote numbers as 'infectious units' - the smallest unit which causes a detectable effect on a susceptible host. A variety of methods are available of which the plaque assay is very popular.

plaque forming units

Plaques are essentially clearings or 'windows' against a background of confluent growth of host cells. Where assays are performed using plaque formation as the measure of infection the term plaque forming unit (PFU) is conventionally used.

When estimating the numbers of virus particles which infect bacterial cells the technique is to mix bacterial cells and virus particles and spread them very thinly as an agar overlay on an agar plate. It is important that the relative concentrations are correct as there must be far fewer virions than host cells.

∏ Why must the number of host cells far exceed the number of virions?

We know that it is possible for many virions to infect a single host cell. Ideally in this technique we want a situation where one virion infects a single host cell giving rise to one plaque.

If none of the virions are infective then a lawn plate showing even, confluent bacterial growth would occur. Where an infection has occurred with subsequent release of virions and further reinfection, a localised area of lysis occurs, that is the formation of a plaque, indicating the presence of one infective unit.

Similar plaque counts can be carried out for animal cell viruses - the cells are grown in a monolayer which is overlayed by a viral suspension.

One other use for experiments involving plaque formation is that the technique can be considered analagous to streak plating for the purification of micro-organisms. It is reasonably safe to assume that one plaque was produced by the progeny of a single virion thus a pure culture has been initiated.

Not all viruses cause cell lysis. One of the most popular methods of estimating some animal viruses such as the 'flu' virus is based on their tendency to adsorb to the surface of red blood cells. If the ratio of viruses to red blood cells is large enough then the viruses will join the blood cells together causing agglutination, hence the technique is a haemagglutination assay. In practice several concentrations of viruses are mixed with the red blood cells to find the minimum concentration which will cause agglutination. Once determined the value can be compared to previously prepared standards and the actual virus particle concentrations calculated.

The use of antibodies to estimate viruses is becoming increasingly important. In this, an antibody which specifically binds with a viral component is used. The technique depends on measuring the amount of antibody that becomes bound. A useful technique is to attach an enzyme to the antibody and measure the amount of enzyme that is bound. There are many forms of this type of assay usually referred to as ELISA (Enzyme Linked Immune Sorbent Assays). They are quick to perform and sensitive. The key to their success depends upon the specificity of the antibody. The use of relatively non-specific antibodies could lead to erroneous results.

Plant virions may cause necrotic lesions of leaves which are distinguishable enough from normal tissue to be counted. Finally some viruses which induce tumours may cause host cells to grow faster giving not plaques but increased 'clumps' of cells, each called a focus of infection.

particle counts and PFUs not always consistent

One phenomenon which takes some time to get used to is the fact that the infectious unit count is generally many times lower than the electron microscope particle count. This ratio between the two values gives rise to the efficiency of plating. The efficiency with which particles infect hosts is thus less than 100%. For most bacterial viruses the figure is usually at or above 50% but in animal viruses the values are much lower, often as low as 0.01%. Many viruses have probably simply been unsuccessful in attempting to infect their host during a particular experiment but the majority apparently do not contain the exact surface sites required to make contact with the specific host cell.

A one-step experiment was carried out giving a curve of similar shape to that shown in Figure 8.8.

From the information given below calculate:

1) the initial number of host cells;

2) the number of virions estimated by electron microscopy;

3) the number of virions estimated by a plaque assay;

4) the efficiency of plating;

5) the number of virions released per host cell.

A viable bacterial count carried out immediately before addition of virions gave 71 cells in 0.1 ml of a 10^3 dilution. 10 ml of this suspension was added to 10 ml of virion suspension and the mixture diluted after two minutes to a total volume of 100 ml. After incubation for a further 40 minutes the number of virions present was determined in the following ways.

A suspension of 250 nm diameter latex beads (10^7 per ml) were mixed with an equal volume of virions and the mixture examined under the electron microscope. There was a ratio of five beads to 19 virions.

After carrying out a plaque assay, 0.1 ml of a 10^5 dilution gave 21 plaques. (Assume all host bacterial cells were infected by phage for the calculation of the number of virions released per host cell).

8.10 Virus purification

Virions are surprisingly stable when confronted with a variety of treatments which would damage or destroy other structures. Techniques such as high speed centrifugation using high centrifugal (g) forces, ammonium sulphate fractionation and solvent treatment may be used to purify certain viruses.

After an experiment such as a large scale one-step reaction a suspension would probably contain a high concentration of virions together with much host cell debris and a few intact whole cells.

The first step is often a short, low speed centrifugation step which will remove whole cells and large pieces of cell debris.

From this point any one of several techniques could be employed to further purify the virions, for example:

Ammonium sulphate precipitation

In this method a carefully measured amount of ammonium sulphate is added so that proteins will be precipitated (hopefully most of them) but not the virus protein. After removing the precipitated protein by centrifugation, addition of a little more ammonium sulphate will precipitate the virion protein. Centrifugation of this precipitate and resuspension of the virus protein pellet will often give a high degree of purification in one step.

Density gradient centrifugation

Density gradient centrifugation is most commonly carried out with gradients of sucrose or cesium chloride. Depending on the centrifugal force and shape of the gradient particles may be separated according to their size or density or both. When properly researched it is possible to design a protocol giving very pure fractions of virus particles.

Solvent extraction

In this technique lipids are solubilised in the organic phase; debris often settles out at the interface and the unaltered virus stays in the aqueous phase.

8.11 The potential of viruses for genetic engineering

Due to the genetic simplicity of virions and their great differences from the host cells which they infect it has quickly become possible to study their genetic systems and processes. The information derived offered opportunities to manipulate DNA within cells, literally to suppress genes or to add new DNA which then became a stable part of the host genome. Examples of the great strides forward made possible using virus research include our understanding of DNA replication, particularly of plasmids capable of independent replication; our understanding of the control exercised by temperate phages during lysogeny and finally the discovery of reverse transcriptase - an enzyme in retroviruses which has given us the means of transcribing information from mRNA back into DNA *in vitro*.

A major thrust in investigative research involves a technique called gene cloning, a method of interfering with the genome to obtain large quantities of specified genes. This process is carried out using cloning vectors in bacteria and largely with the aid of bacteriophages. Thus we can insert 'foreign' genes into bacteriophage nucleic acid and use the lytic cycle to produce large quantities of these genes. Alternatively we may put a 'foreign' gene into the nucleic acid of a temperate phage. After infection, the foreign gene (now part of a prophage) may be expressed in the new host. This is one of the core techniques of genetic engineering. A bacteriophage used in this way is described as a vector. Similar vector systems using viruses have been developed for animal and plant systems. The details of using these systems as genetic vectors are provided in the BIOTOL texts 'Techniques for Engineering Genes' and 'Strategies for Engineering Organisms'.

So far genetic engineering has helped in several areas which affect us directly or indirectly. For example the increased production of medically essential compounds such as hormones, blood clotting factor VIII and interleukin-2 have already or are being developed. Several other possibilities are being considered from developing organisms to produce desirable quantities of enzymes, to alter yeasts to further optimise production in the brewing and fermentation industries and to develop organisms which degrade recalcitrant compounds more quickly. Resistance to disease in plants can be genetically engineered - a process giving an alternative to the sometimes indiscriminate use of pesticides - although the engineering itself is not without dangers.

The list of achievements is relatively short as yet but current developments are many and the future could be very rewarding providing we act responsibly.

Summary and objectives

This chapter provides an introduction to virology. It sets out to give an introduction to the history of the subject and the terminology used, the structure and diversity of viruses and an indication of how viruses are classified. Viral replication by both lytic and lysogenic processes was followed in some depth and special emphasis was placed on bacteriophages. At this stage the techniques of cultivation, quantitation and purification of viruses were explored.

Now that you have completed this chapter you should be able to:

- compare and contrast the properties of viruses to those of living cells;

- outline the development of virology as a science indicating its importance to microbiology;

- discuss the classification of viruses based on selected properties;

- recognise and describe the different structural classes of viruses;

- describe the structure and functions of the various components within the virus;

- describe the replication of viruses, either by the lytic cycle or by lysogeny;

- describe the strategy of and events occurring within the one-step experiment to culture viruses;

- develop a protocol for the purification of viruses;

- decide on the most suitable method for the quantitation of viruses.

Chemical control of microbial growth

Chemical control of microbial growth

9.1 Introduction

In Chapter 6 we looked at ways in which microbial growth is inhibited by naturally occurring factors and how some of these may be intentionally applied to kill micro-organisms (eg by heat, irradiation, high solute concentration, etc). There are many instances where these methods of inhibiting microbial growth are impossible or impractical. One reason may be that the material to be freed of microbes is unsuitable (for instance plastics cannot withstand heat) or too large (floors, walls, worktops, etc). In such cases chemical agents may be the answer; of which the so-called disinfectants can be applied to inanimate objects. If microbes in humans, animals and plants, need to be inhibited, that is, if the chemicals are used for therapy, they are called chemotherapeutic agents.

disinfectants and chemotherapeutic agents

Antimicrobial agents can be either non-selective or selective in their action. If an agent selectively kills microbial cells without inhibiting host cells, it may become an important weapon for combating disease. Antibiotics are antimicrobial agents produced by micro-organisms which can be tolerated by most larger eukaryotic organisms. After the discovery of their therapeutic value, the great demand for antibiotics necessitated research into ways of chemically manufacturing them. This was achieved during the early 1950's, and since then antibiotics have found an even wider application in the medical, as well as in the agricultural, field. Yet, antibiotics do have their drawbacks.

There are many cases known in which micro-organisms have developed some form of resistance to antibiotics which had previously inhibited growth. The resistance may take one of several forms and may be temporary or permanent but the consequence is a loss of the therapeutic value of the antibiotic. This phenomenon urges scientists not only to continue their search for new antibiotics, but also to do research into possibilities of altering the chemical composition of known antibiotics in such a way that these agents can still inhibit their target: the infecting micro-organism.

In this chapter we will consider the many aspects of inhibiting microbial growth chemically, the ways in which antimicrobial agents act, the therapeutic value of natural and synthesised antibiotics, and the problems that have risen and been solved during their history.

9.2 Antimicrobial agents

9.2.1 Different types of antimicrobial agents

There are many instances where microbial growth is unwanted and should be either restricted or completely eliminated. Chemicals exhibiting either of these effects are known as antimicrobial agents. A distinction can be made between substances that attack specific cells only, ie chemicals that are selective in their target, and non-selective compounds that damage different kinds of cellular organisms. Such selective agents become of therapeutic value if they inhibit the growth or kill infecting micro-organisms,

selectivity

without harming the host. Non-selective antimicrobial agents are not only harmful to micro-organisms but also the animal and plant tissue. They are usually synthetic chemicals.

Antimicrobial agents can be broadly divided into different types according to how they are used.

antiseptics   Antiseptics are compounds that are used to inhibit or occasionally kill micro-organisms on skin, mucous membranes and other living tissue without harming these surfaces. They are not suitable for internal use within either the digestive system or blood stream. The use of the word 'germ' has declined but antiseptics are still often referred to as germicides, that is, killers of germs.

disinfectants   Disinfectants are chemicals that kill or inhibit micro-organisms, but are unsafe for application on any living tissue. It follows that disinfectants are only applied to reduce or eliminate contamination on inanimate surfaces, such as floors, work and table tops and utensils. Disinfectants are active against the vegetative (growth) stages of bacteria, fungi and viruses, but not all of them kill spores. Those that can are termed sporicidal disinfectants and include chemicals such as gluteraldehyde, formaldehyde and ethylene oxide. These chemicals are not only hazardous to spores but also to humans and great care must be taken when applying them.

selective toxicity   Selective microbial agents are more toxic to microbes than to their host, either plant or animal. Agents that kill disease-causing prokaryotes (bacteria), without harming their eukaryotic host, are of particular medical significance. But also agents that are effective against eukaryotic micro-organisms (algae, fungi, protozoa) and that are harmless to eukaryotic cells of higher organisms are important for medical purposes. Together, these selective microbial agents are known as chemotherapeutic agents. They may be **chemother-apeutic agents** either of natural origin or synthesised. Antibiotics belong to the former group since they are produced by certain micro-organisms though many current antibiotics are modified **antibiotics** natural compounds and a few are now totally synthetic.

9.2.2 Effects of antimicrobial agents

cidal and static agents   Antimicrobial agents either kill microbes or inhibit their growth. The suffix cidal implies that the agent actually kills germs. So, fungicides kill fungi, bacteriocides kill bacteria and algicides kill algae. If the agent only inhibits growth, the suffix static is used: fungistatic, bacteriostatic and algistatic agents inhibit growth of fungi, bacteria and algae, respectively. However, it is not always possible to make a clear-cut distinction between cidal and static agents: a cidal agent may indeed be cidal at high concentration, but static at lower concentrations.

A static agent must be present all the time in order to be effective: as soon as it is removed or its activity neutralised in some way, then any of the viable microbes still present can resume growth.

A knowledge of how a chemical agent affects microbial growth is of great importance in understanding the way it acts. In Chapter 4 we looked at the various stages of growth in a microbial culture. It is the exponential phase of microbial growth that is the most suitable stage in which to study the effect of an antimicrobial agent. If the agent is added to a culture in that particular phase at an inhibitory concentration, and samples are continuously taken for measurement of growth and viability, an accurate picture is obtained of the way the agent acts (Figure 9.1). It is also during the exponential phase of growth the point at which micro-organisms are most sensitive to antimicrobial agents.

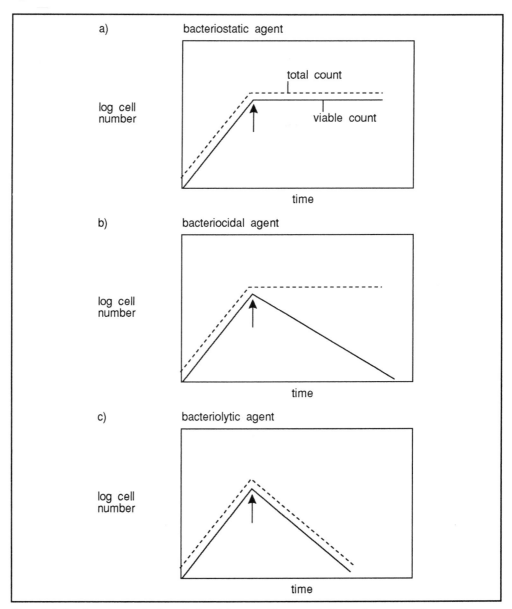

Figure 9.1 The three types of action of microbial agents. The arrow indicates the moment the agent is added to the culture. Note the relationships between total cell count and viable count.

The effect of the agent on the exponentially growing bacteria culture may be either bacteriostatic, bacteriocidal, or bacteriolytic. As we noted before bacteriostatic agents do not kill the micro-organism but only inhibit growth (Figure 9.1a). This is often caused by inhibition of protein synthesis, as the agent binds to the ribosomes. This binding, however, is reversible and when the agent's concentration is lowered growth is resumed. Bacteriocidal agents prevent growth by killing microbial cells but without rupture or lysis (Figure 9.1b). They bind to certain specific targets in the cell, but, in contrast with bacteriostatic agents, their effects are irreversible.

lytic agents Bacteriolytic agents induce cell lysis which can be measured by cell counting or turbidity (Figure 9.1c). Such agents act for instance, by inhibiting cell wall synthesis (antibiotics, like penicillin), or they may act on the cell membrane.

∏ We can see from Figure 9.1 that the effect of an antibacterial agent can be distinguished by comparing total and viable cell counts. Now consider practical aspects of this experiment and explain, in a few words, how you would 1) estimate the total number of cells and 2) estimate the number of viable cells.

1) The total number of cells may be estimated by a variety of methods. For instance, by direct microscopic counting or by measuring the turbidity of the suspension.

2) The most suitable method for estimating the number of viable cells is by colony counting after transfer of suitable dilutions of the suspension to a solid recovery (growth) medium. An important point is that the antibacterial agent must not inhibit the growth on the recovery medium. In some cases dilution of the agent is sufficient to eliminate its growth inhibitory effects. In other instances the agent may have to be neutralised by chemical means or the organism may have to be separated from the agent by filtration or centrifugation.

9.2.3 Measuring antimicrobial activity

We have seen that antimicrobial agents can be either bacteriostatic, bacteriocidal or bacteriolytic in their effect. Another criterion for the applicability of an antimicrobial agent is the rate at which the micro-organisms are killed. This can be estimated by determining the number of viable organisms at certain intervals after having mixed a suspension of micro-organisms with the agent. An example is given in Figure 9.2, in which the number of surviving spores of *Bacillus* is plotted against the time of exposure to the disinfectant.

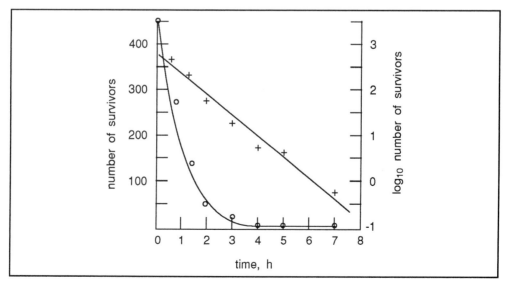

Figure 9.2 Survival of spores in a *Bacillus* suspension after the addition of a certain disinfectant.
o - o = number of survivors, + - + = log₁₀ number of survivors.

time-survivor
relationship Such a time-survivor relationship often shows an exponential curve. This means that
the rate at which the cells are killed only depends on the number of viable cells at a
certain moment. Mathematically this can be expressed as follows:

$$\frac{dN}{dt} = -kN$$

in which:

$\frac{dN}{dt}$ = the decrease in number of viable cells in the time interval dt.

k = killing constant.
N = number of viable cells.

and after integration of the equation:
$N_t = N_o e^{-kt}$,

in which:
N_o = number of viable cells at time zero.
N_t = number of viable cells after exposure to antimicrobial agent for time interval t.

The killing constant k has a different value for each micro-organism and is also
dependent on the micro-organism's physiological state and several environmental
factors. Of these temperature, pH, and chemical composition of the medium are the
most important. Since the time-survivor curve is usually exponential the activity of the
decimal
reduction time antimicrobial agent may also be expressed as the decimal reduction time or D_{10}-value.
This is the time necessary to decrease the number of viable cells by a factor of 10.

When chemically controlling microbial growth the concentration of the antimicrobial
agent is also of great importance. This is shown in Figure 9.3 for the effect of various
concentrations of phenol on *Escherichia coli*.

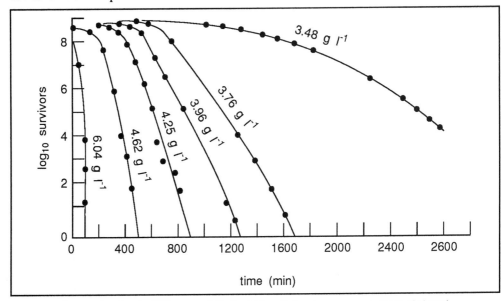

Figure 9.3 Curves showing the killing rate of *Escherichia coli* at various concentrations of phenol.

∏ Can you think why phenol, which denatures proteins, has a plateau region in its kill curve against *E. coli* (Figure 9.3)?

The plateau is explained by the requirement for damage to many proteins within the same cell before death occurs ie denaturation (inactivation) of a single enzyme would not cause death but inactivation of a number of enzymes would. This is a multi-hit model rather than a single-hit model.

∏ Can you think what could account for the shapes of the kill curves presented below.

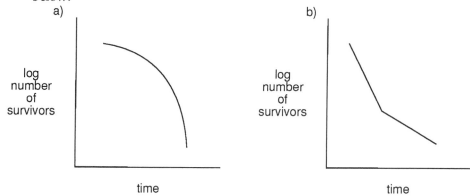

a) Cell clumping or the presence of debris that protects the organism from the action of the disinfectant.

b) The presence of two types of cells with different killing constants.

SAQ 9.1

Examine Figure 9.2:

1) Determine a D_{10}-value for the disinfectant from these results.

2) Select, from the list provided factors that would increase the rate of killing by a disinfectant.

    a) Cell clumping.

    b) A higher concentration of disinfectant.

    c) An extended time of exposure.

    d) A lower number of cells.

    e) A lower killing constant.

MIC

The antimicrobial activity of an agent can be measured by the minimum amount needed to inhibit growth of a test organism, the so-called minimum inhibitory concentration (MIC). The MIC is not a constant with a typical value for an antimicrobial agent since it varies with the circumstances: the composition of the culture medium, the test organism used, the inoculum size, the incubation time and the conditions during incubation, such as pH, temperature and aeration. If agents need to be compared to find the most effective agent against a certain organism, strict control of these circumstances are necessary.

tube dilution technique

There are two methods of determining the MIC of an agent in common practice: the tube dilution technique and the agar diffusion method. In the tube dilution technique a series of test tubes is filled with a growth medium, each tube containing a different concentration of the antimicrobial agent. The tubes are then inoculated with the same number test organisms. After incubation, the turbidity of the medium is an indication of the effect of the agent: the minimum inhibitory concentration is the lowest concentration that causes no turbidity (Figure 9.4).

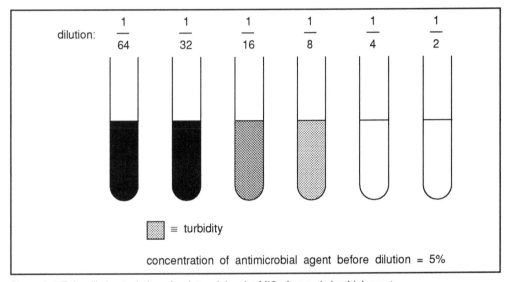

Figure 9.4 Tube dilution technique for determining the MIC of an antimicrobial agent.

Π Examine Figure 9.4 and determine the MIC of the antimicrobial agent.

We can see from Figure 9.4 that the highest dilution (lowest concentration) that results in no turbidity is 1 in 4. The concentration of the agent before dilution is given as 5%, so the MIC is $\frac{5}{4} = 1.25\%$.

It is important to note that the tube dilution method does not distinguish between the way the micro-organism's growth is inhibited, ie between a cidal or a static effect of the agent, since the agent is present during the entire incubation time.

Π Describe what you would have to do to determine the minimum killing concentration (MKC) using the tubes shown in Figure 9.4.

The MKC could be determined by subculturing from the tubes showing no turbidity into fresh growth medium with no added disinfectant. High dilution on subculturing would eliminate 'carry-over' antimicrobial action and any surviving viable cell would then be able to grow in the new growth medium. After incubation of these tubes the MKC is determined by examining for turbidity, as for the MIC determination.

The agar diffusion method may be used to estimate either the concentration or the MIC values of an antimicrobial agent. Several variations of the same basic method are available; probably the most often used being the Kirby Bauer test. In this method a petri dish with agar medium is inoculated with the test micro-organism, either by pouring a layer of agar containing the organism over it, or by spreading a broth culture of the organism over the surface of the agar. Filter-paper discs containing a known quantity of the antimicrobial agent are then placed on the surface of the agar. During incubation the agent diffuses into the agar creating a concentration gradient around the disc. At a certain distance from the paper the concentration becomes the minimum inhibitory concentration, beyond which growth can be seen. The size of this so-called zone of inhibition can be measured with a ruler, and is related to the amount of agent that was originally present in the paper disc (Figure 9.5).

Kirby Bauer test

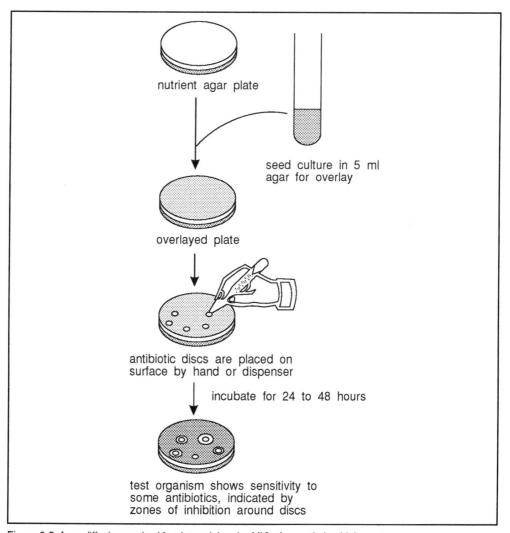

nutrient agar plate

seed culture in 5 ml agar for overlay

overlayed plate

antibiotic discs are placed on surface by hand or dispenser

incubate for 24 to 48 hours

test organism shows sensitivity to some antibiotics, indicated by zones of inhibition around discs

Figure 9.5 Agar diffusion method for determining the MIC of an antimicrobial agent.

∏ Does the agar diffusion method distinguish between cidal and static effects of the antimicrobial agent? Give a reason for your response.

No, because the agent is present during the entire incubation time.

∏ Can you think why the diameter of the zone of inhibition obtained is directly proportional to the log concentration of antimicrobial agent?

The explanation for this is that the dilution of antimicrobial agent in the growth medium increases exponentially as the distance from the disc increases.

SAQ 9.2

The petri dish presented below shows the results of an agar diffusion experiment designed to determine the concentration of the antimicrobial agent in solution Y.

		concentration (μg ml^{-1})	Log$_2$ concentration
standard	S1	1.27	0.35
	S2	4.29	2.10
	S3	13.93	3.80

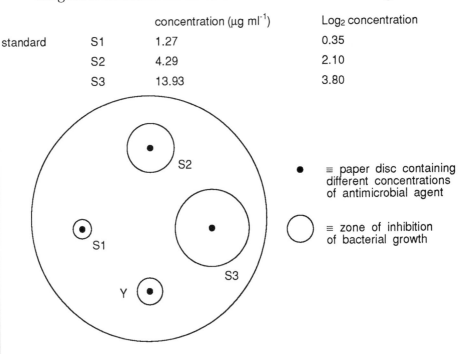

• ≡ paper disc containing different concentrations of antimicrobial agent

○ ≡ zone of inhibition of bacterial growth

Plot a suitable graph and determine the concentration of solution Y.

Let us now look at the way in which the MIC values can be determined using the Kirby Bauer test. It is first necessary to obtain MIC values (for example by the tube dilution method) and zone diameters for many different microbial strains in order to produce calibration curves. If a plot of MIC (log$_2$) against zone of inhibition diameter (arithmetic scale) is produced for a particular anitmicrobial agent a linear relationship is observed between an increase in diameter and a decrease in MIC.

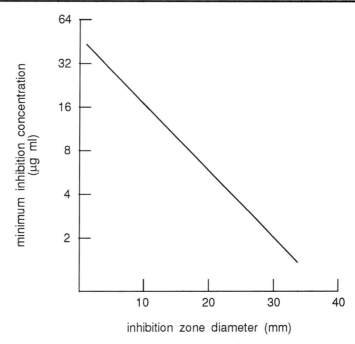

inhibition zone diameter (mm)

Our answer in SAQ 9.2 gave a zone diameter which indicated a concentration of 2 µg ml[-1] but we can now deduce from the graph above that the MIC is approximately 20 µg ml[-1]. This second piece of information can be used to tell us if the agent would be clinically useful.

∏ If we know that the achievable blood levels of the antimicrobial agent used in SAQ 9.2 were between 10 and 30 µg ml during normal treatment and the MIC of that agent against an infective organism was 20 µg ml[-1] - could the agent be recommended for clincial use in this case?

The answer is yes because for most of the time the blood level of the agent exceeds the required MIC to kill the organism.

∏ Could we adjust the dosage regime to improve the treatment?

If it were possible the dosage concentration could be increased or the number of doses per day increased to maintain the blood levels above the MIC at all times.

9.3 Antiseptics and disinfectants

In Section 9.2 we have made a distinction between antiseptics and disinfectants: antiseptics can be applied to the skin, whereas disinfectants are only used on inanimate surfaces. As we indicated disinfectants are more toxic than antiseptics but they are also more active. This relationship is very common with antimicrobial agents, the more active a compound the more toxic it is.

Although one tends to think of antiseptics and disinfectants as belonging to the medical world they are also widely used in industry. The aim of application of germicides in this field is usually to prevent deterioration of materials by micro-organisms.

Generally disinfectants are suitable antimicrobial agents in circumstances where it is impractical to use heat for example certain instruments and equipment used in medicine and in the food industry, on floors, walls, surfaces of work tops and large equipment, etc. Also, nowadays, drinking water is often disinfected, as is wastewater before it is discharged into the environment. An important point to remember is that treatment with disinfectants does not necessarily lead to sterile conditions (an object is called sterile if there are no living organisms present). In most cases disinfection only results in a considerable reduction of the microbial load, and, hopefully, in complete elimination of the pathogenic micro-organisms. For some pathogens this is often not the case: many bacterial endospores and some vegetative cells, such as *Mycobacterium tuberculosis* (which causes tuberculosis), are very resistant to many antimicrobial agents. Table 9.1 summarises the properties of the various classes of antiseptics and disinfectants. You should familiarise yourself with the properties and use of at least one antimicrobial agent from each class.

Heavy metals:	
Mercury: $HgCl_2$	Combines with SH in proteins; very toxic to humans; activity neutralised by SH compounds and organic matter; used as disinfectant
Silver: $AgNO_3$	Protein precipitant; activity neutralised by organic matter; may be used in eyes of newborns to prevent gonorrhoea
Copper: $CuSO_4$	Algicide in swimming pools and water supplies; fungicide for plant diseases
Halogens:	
Iodine: Tincture of iodine (0.2% I_2 in 70% ethanol), iodophors (I_2 complexes with detergents)	Iodinates proteins containing tyrosine residues; antiseptic on skin; disinfectant of medical instruments; small-scale water purification; relatively nontoxic on skin but toxic internally
* Chlorine: Cl_2 gas, $Ca(OCl)_2$, $NaOCl$; active ingredient is $HOCl$, formed at neutral or in acidic conditions	Oxidising agent; activity neutralised by organic matter and NH_3; used in water purification, general disinfectant in food and dairy industries
* **Alcohols:**	
Ethanol, isopropanol; most active with some water present (50-70% alcohol in water)	Lipid solvents and protein denaturants; antiseptic on skin; disinfectant for hospital items
Aldehydes:	
Formaldehyde, HCHO; available as 37% solution (formalin); used as 2% aqueous solution	Alkylating agent; combines with NH_2, COOH, and SH groups in nucleic acids and proteins; neutralised by organic matter; used for embalming of corpses; small amounts in wood smoke participating in meat-smoking process

Table 9.1 Classes of antiseptics and disinfectants and their uses. Adapted from Brock T.D; Smith, DW and Madigan, MT 1984, Biology of Micro-organisms, Prentice Hall.

Glutaraldehyde, $OHCCH_2CH_2CH_2CHO$; used as 2% aqueous solution	Less toxic than formaldehyde; cold sterilisation of hospital goods; neutralised by organic matter
Ethylene oxide: Used as a gas $\overset{O}{\overset{/\backslash}{CH_2 - CH_2}}$	Alkylating agent; very toxic, used in special gas sterilisation units for heat-sensitive goods
* Phenols: Phenols, carbolic acid, C_6H_5OH; used as aqueous solution (5%); Many phenol derivatives used: with halogen, alkyl, OH groups (cresols, thymol, etc)	Protein denaturant; disrupts cell membrane at low concentration; very toxic, activity increased by soaps, not affected by organic matter; disinfectants for large, dirty surfaces (floors, walls, etc)
* Cationic disinfectants: where one of the R groups is a long-chain alkyl and the other three are methyl, benzyl, etc $\overset{R'}{\underset{Cl^-}{R - N^+ - R''}}$ R'''	Affect cell membrane through charge interactions with phospholipids; neutralised by phospholipids; metal ions, low pH, soaps, organic materials; skin antiseptics; disinfectants for medical instruments, food and dairy equipment
Ozone, O_3: gas; must be generated at the site of use (high-voltage electric discharge)	Oxidising agent; very toxic to humans; water purification; action neutralised by organic matter

Table 9.1 (cont'd) Classes of antiseptics and disinfectants and their uses.

We will now consider some of most widely used antiseptics and disinfectants in more detail - these are indicated by an asterisk in Table 9.1.

Alcohols as antiseptics

organic alcohols

The organic alcohols ethanol, isopropanol and benzyl alcohol are effective antiseptics as 50 to 70% aqueous solutions. They precipitate (denature) proteins and solubilise the lipids of the cell walls and cell membranes of bacteria. An alcohol can be used to disinfect the human skin, either on its own or in combination with iodine. Alcohols are effective killers of vegetative bacterial cells, and fungi but they are ineffective against spores. Also, they do not inactivate all viruses.

Chlorine as disinfectant

chlorine

The halogen chlorine is used to decontaminate, for instance, swimming pools and main water supplies. Its activity is neutralised by the presence of organic material. Chlorine reacts with water, forming hypochlorous acid (HOCl):

$$Cl_2 + H_2O \rightarrow HCl + HOCl$$

Hypochlorous acid reacts with water forming HCl and peroxide (H_2O_2). Both products, hypochlorous acid as well as hydrogen peroxide, are strong oxidants, and toxic to biological systems.

$$HOCl + H_2O \rightarrow HCl + H_2O_2$$

Hypochlorous acid is probably best known in the form of household bleach in which sodium hypochlorite occurs in a 5.25% solution.

A 10% solution of household bleach in water is an effective disinfectant, as long as the surfaces can withstand its action. For the disinfection of water supplies dry bleach ($Ca(OCl)_2$) is used which liberates free chlorine gas on addition to water.

∏ Can you think of a practical advantage of using dry bleach as apposed to chlorine gas for the disinfection of water supplies.

Because dry bleach is a solid it is more easily transported in bulk than chlorine gas.

∏ Give a reason why chlorine is suitable for the disinfection of water but cannot be used as a soil disinfectant.

It is inactivated by organic material.

Phenols as disinfectants

phenol   Phenol (C_6H_5OH) and its substituted derivatives (phenols) are very effective as disinfectants. Phenol was used by Joseph Lister in 1867 to prevent infection after surgery. Since it is very caustic to human tissue phenol has been replaced by several substituted derivatives. Certain phenol derivatives are a good example of germicides that can be used as antiseptics at low concentration and as disinfectants at higher concentration. They denature proteins and disrupt cell membranes. There are several phenol derivatives on the market (Lysol, Clearsol, Dettol) that are of varying effectiveness against many bacteria, and that are not readily inactivated by organic compounds. They are especially useful as disinfectants in the laboratory.

hexachlorophene   Hexachlorophene is a very powerful chlorinated phenol. It kills staphylococci at concentrations as low as one part per million. It used to be applied as an antiseptic but, since it has been proved to be readily absorbed by the skin and to cause brain damage in laboratory animals, it can now only be bought on prescription.

Cationic compounds as disinfectants

Cationic compounds along with certain anionic and amphoteric substances are surface active agents, that is they lower the surface tension of water resulting in solutions or suspension which wet out available surfaces more readily. These compounds generally have cleansing or detergent properties such as a kitchen surface which is contaminated by grease. A compound having detergent and disinfectant properties will penetrate the grease layer and remove it leaving a clean layer which is disinfected at the same time. A non-detergent may simply slide over the grease contamination leaving any entrapped micro-organisms protected and unaffected.

Cationic compounds are so called because the active part of the molecule in terms of disinfectant activity resides in the cation or positive ion. Such compounds are intrinsically more effective than anionic disinfectants. This is because all living cells carry an overall negative charge (though this is made up of many positive and many negative groups on their surfaces). Thus there will be attraction between the negatively charged cell and the active positive ion of the disinfectant.

A major advantage of cationic disinfectants is that they are relatively non-toxic and non-injurious to humans and when used as recommended are very safe in hospital and food processing environments.

One disadvantage of these compounds however is that they are adversely affected by the presence of hard water (water containing Ca^{2+} ions).

A prominent example of a cationic disinfectant is the group of quaternary ammonium compounds (QACs) which have a pentavalent nitrogen attached to one or more hydrophobic chains and of the general formula:

$$R_4.N^+ \ X^-$$

Due to their low toxicity, cleansing properties and disinfectant activity cationic disinfectants are used extensively in the disinfection of instruments, food utensils, surfaces and as skin antiseptics.

Disinfectant testing

As we saw in the previous section the determination of the MIC of a disinfectant in the laboratory does not pose any particular problems. However, this does not necessarily reflect antimicrobial effectiveness in use. There are several reasons for this. First of all the disinfectant may be neutralised by organic materials, so that the effective concentration is not maintained for a long enough period. Secondly, micro-organisms are often encased in particles, which hamper penetration of the agent to the cells. So, the effectiveness of a disinfectant should be determined for the conditions in which it is to be used. For example, an in-use test for a surface disinfectant could involve recovery of viable organisms, by surface swabbing, before and after application of a disinfectant.

phenol-
coefficient

For more quantitative data the effectiveness of a disinfectant can be evaluated by standard procedures used by manufacturers of disinfectants. One approach is the determination of the phenol-coefficient. Here, the efficacy of a phenolic-type disinfectant is compared with that of phenol under standard conditions. There are many tests to determine the phenol coefficient. Generally, a dilution series is made of the disinfectant and a certain amount of each dilution is added to tubes with a fixed amount of bacterial suspension which are then left for certain lengths of time (for instance 2.5, 5.0, 7.5 and 10 minutes). Next, a loopful from each tube of exposed bacterial suspension is transferred to a broth and left to incubate. Broth cultures showing growth are scored positive, those without growth negative. The scores are set against those of an identical test carried out with phenol.

The phenol-coefficient is the ratio of the minimum sterilising concentration of phenol: to the minimum sterilising concentration of test disinfectant which kills the culture in 7.5 minutes. The test as described is the Rideal-Walker test.

∏ If a disinfectant has a phenol-coefficient greater than unity, is it more effective or less effective than phenol?

The disinfectant is more effective than phenol.

∏ The phenol-coefficients of four phenolic type of disinfectants are given:

Disinfectant	Phenol-coefficient (Rideal-Walker)
A	7.0
B	0.3
C	1.8
D	2.1

1) List the disinfectants in decreasing order of antibacterial effectiveness; 2) Which of the disinfectants are more effective than phenol itself?

You should have come to the following conclusions.

1) The higher the phenol-coefficient the more effective the disinfectant under the conditions of the Rideal-Walker test. The decreasing order of effectiveness is therefore A, D, C, B.

2) Phenol has a phenol-coefficient of 1.0. Disinfectants A, C and D are therefore more effective than phenol.

Rideal-Walker test

The phenol coefficient test is an example of a suspension test, in other words a suspension of micro-organisms is mixed with a disinfectant and after certain exposure times the presence of or actual number of survivors is determined. The phenol coefficient or Rideal-Walker test, was the first widely used disinfectant test; introduced by Rideal and Walker in 1908 and achieving a British Standard in 1934. It is only applicable to phenolic compounds and is qualitative in that survivors are not counted, readings are merely positive or negative.

Kelsey Sykes test

Since 1908 many different test methods have been published with a tendency to produce a specific test for a specific end use. For instance the Kelsey Sykes test (1967) is an 'in-use' suspension test designed to determine recommended disinfectant concentrations for use in hospital situations.

QAC test

There is a test specific for quaternary ammonium compounds based on finding the activity of a QAC against a specific *E. coli* strain. It is quantitative, measuring actual numbers of survivors and is intended to evaluate compounds for domestic and industrial but not hospital use.

As an on going European initiative to produce a standard, internationally recognised test a '5-5-5' test based on a Dutch suspension test is being considered. The title comes from the use of five standard organisms, a five minutes challenge period and the criterion of activity being a reduction in viable cell numbers of five log cycles.

| SAQ 9.3 | For each of the disinfectants 1 to 5, select appropriate characteristics from the list provided (a to h). |

Disinfectant

1) Phenol.

2) Chlorine.

3) Ethanol.

4) Cationic detergents.

5) Ethylene oxide.

Characteristics

a) Inactivated by organic matter.

b) Commonly used as an antiseptic.

c) Denatures proteins.

d) Oxidising agent.

e) Neutralised by phospholipids.

f) Alkylating agent.

g) Ineffective against spores and some viruses.

h) May be used as a gas.

9.4 Chemotherapeutic agents and their mode of action

9.4.1 Selectivity of antimicrobial agents

selectivity   In Section 9.2 we made a distinction between selective and non-selective antimicrobial agents. Disinfectants and antiseptics belong to the latter group and chemotherapeutic agents to the former. The selectivity of a chemotherapeutic agent (drug) is based on the presence of a unique structure or process in the micro-organism which is absent in the host. Such sites and processes are the bacterial cell wall, cell membrane, ribosomes, and protein and nucleic acid synthesis. There are two categories of chemotherapeutic agents:

1) Antibiotics. These metabolites are produced by micro-organisms in low concentration and are toxic to other microbes. The toxicity is probably of survival value to the producing micro-organism. There are many antibiotics known but only a few can be applied in medicine since the majority are too toxic for humans. As we shall see later in this chapter, these days many antibiotics are in fact (partly) chemically synthesised.

2)   Chemosynthetic agents. These are man-made compounds that are selectively toxic to micro-organisms.

Several factors determine whether an agent is an effective drug. Firstly, of course, the agent should inhibit or kill micro-organisms. But, apart from that, its stability *in vivo*, rate of absorption, rate of elimination and rate of penetration into the body site to be treated all contribute to the efficacy of the drug. If a drug fails in any one of these characteristics there are often ways of overcoming the problem, sometimes by chemical modification, sometimes by administering via a different route: for instance, by injection instead of oral administration if the drug is not absorbed via the digestive tract.

therapeutic index

An important characteristic of a drug is its therapeutic index which is defined as the minimum dose which is toxic to the host divided by its effective therapeutic dose. A high index means a high efficacy. All factors contributing to the efficacy affect the therapeutic index and the drug's efficacy is thus the result of a complex interaction between patient, drug and micro-organism.

9.4.2 Production of antibiotics and their activity spectrum

the discovery of penicillin

Like so many important discoveries that of antibiotics happened by chance. The Scottish physician Alexander Fleming was doing research on *Staphylococcus* strains when he accidentally discovered penicillin. He necessarily had to open the petri dishes for examination which, unavoidably, caused some contamination of the culture plates by airborne micro-organisms. After some time Fleming noticed a transparent zone in the *Staphylococcus* colonies around a large colony of a contaminating mould. He assumed that a chemical produced by the mould had passed into the solid medium and inhibited the growth of staphylococci. He called the active agent penicillin and published his findings in 1929. Fleming was unable to purify the agent, but still saw and advocated its medical application. This, however, did not take place until the early '40's when Howard Florey and Ernst Chain managed to purify a sufficiently large quantity of penicillin and demonstrated its effectiveness against a streptococcal infection in mice. Florey and Chain then began to collect larger quantities of the antibiotic, and, in 1941, when they thought they had collected enough penicillin, they treated a human patient for a bacterial infection. However, after initial improvement, the patient died because they ran out of the drug!

The next important step in the development of penicillin had to be large-scale production. This was finally achieved in the US by American and British scientists, so that, by the end of World War II penicillin could be applied to cure infectious diseases on a larger scale.

6-APA

Finally, large-scale production was boosted in 1959, when 6-amino penicillanic acid (6-APA) was isolated in large quantity. This intermediate product in the biosynthesis of natural penicillin can serve as starting material for many semisynthetic penicillins.

6-aminopenicillanic acid

The success-story of penicillin urged many scientists to look for other micro-organisms in nature that produce antibiotics. However, this ability is limited to only a few genera of fungi and bacteria, the main ones are listed in Table 9.2.

Chemical Class	Generic names	Biological Source
β-lactams	Penicillins Cephalosporins	*Penicillium spp.* *Cephalosporium spp.*
Macrolides	Erythromycin Carbomycin	*Streptomyces erythreus* *Streptomyces halstidii*
Aminoglycosides	Streptomycin Neomycin	*Streptomyces griseus* *Streptomyces fradiae*
Tetracyclines	Tetracycline[a]	*Streptomyces aureofaciens*
Polypeptides	Polymyxin g Bacitracin	*Bacillus polymyxa* *Bacillus subtilis*
Polyenes	Amphotericin B Nystatin	*Streptomyces nodosus* *Streptomyces nouresii*
	Chloramphenicol[b]	*Streptomyces Venezuelae*

Table 9.2 Biological sources of certain antibiotics.
[a] Made microbiologically and by chemical dehydrochlorination of chlorotetracycline.
[b] Now made by chemical synthesis.

Streptomyces   We can see from Table 9.2 that members of the genus *Streptomyces* are especially important as antibiotic-producing bacteria. In fact, they provide more than 50% of the known antibiotics. It is assumed that the majority of antibiotic producing micro-organisms have been discovered and research is now aimed at improving the properties (and thus the therapeutic value) of the known antibiotics.

Any antibiotic has a limited range of bacteria against which it is effective. This range is known as the antibiotic's spectrum and the spectrum can be either broad or narrow.

broad-spectrum   Broad-spectrum antibiotics inhibit a wide range of both Gram-negative and Gram-positive bacteria. The tetracycline group forms a good example of broad-spectrum antibiotics: tetracyclines inhibit Gram-positive, Gram-negative, aerobic and a few anaerobic bacteria, some protozoa and several other micro-organisms. Chloramphenicol and cephalosporins also have a broad spectrum.

narrow-spectrum   Narrow-spectrum antibiotics inhibit either Gram-negative or Gram-positive bacteria, eg most penicillins are effective only against Gram-positive bacteria.

9.4.3 Mode of action of antibiotics

The antimicrobial properties of an antibiotic are related to its mode of action. A knowledge of how antibiotics inhibit or kill microbes is therefore important for establishing how the antibiotic should be used.

Four different mechanisms can be distinguished in the ways antibiotics affect micro-organisms:

- inhibition of cell wall synthesis;

- influence on structure and permeability of the cell membrane;

- inhibition of protein synthesis;

- inhibition of nucleic acid synthesis.

These mechanisms are illustrated in Figure 9.6 and will be discussed briefly.

∏ It would be a good idea to transfer the information shown in Figure 9.6 into a table. We suggest the following format.

Mechanism	Antibiotics
Inhibition of cell wall synthesis	Pencillins

You may add to your list as you learn more from your studies.

Antibiotics that inhibit cell wall synthesis

The main function of the microbial cell wall is to give the cell rigidity and protection. If the cell loses its cell wall by chemical disruption or if synthesis is inhibited, it will be lysed because it has become osmotically sensitive. The antibiotics either inhibit enzymes in cell wall synthesis or activate bacterial enzymes, known as autolysins, that disrupt the cell wall.

penicillins and cephalosporins Penicillins and cephalosporins are groups of antibiotics that inhibit cell wall synthesis. Penicillin G, which is the natural antibiotic of *Penicillium notatum* and *P. chrysogenum*, inhibits the final step in the synthesis of the peptidoglycan of the bacterial cell wall. This final step is the formation of the peptide cross-links between adjacent glycan chains, known as transpeptidation or cross-linking (see Figure 9.7). As a result of the action of penicillin peptidoglycan is formed which lacks strength so that the cell eventually lyses.

Most penicillins have a narrow spectrum. They are effective against Gram-positive bacteria, but have hardly any effect on the majority of Gram-negative bacteria, since they can not penetrate the outer membrane of the cell wall. They are also not active against non-growing cells. Penicillins have a high therapeutic index and they are particularly suitable for treatment of most cases of gonococcal and streptococcal infections.

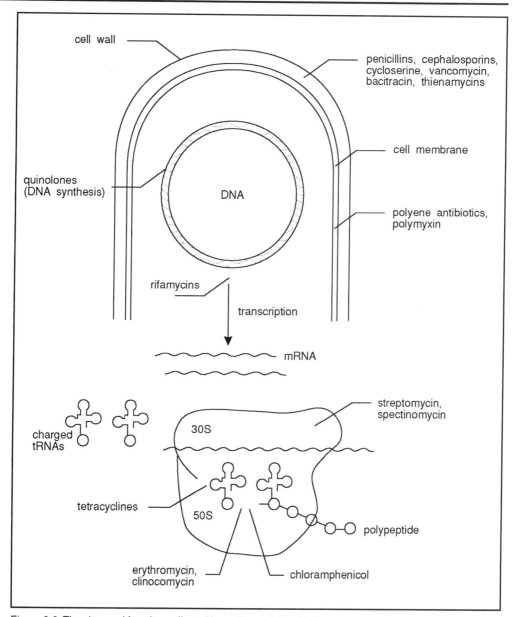

Figure 9.6 The sites and functions affected by antibacterial antibiotics.

∏   Can you think why penicillins are not active against non-growing cells?

Penicillins interfere with the synthesis of the cell wall. New constituents of cell walls are only formed in growing cells, not in resting cells.

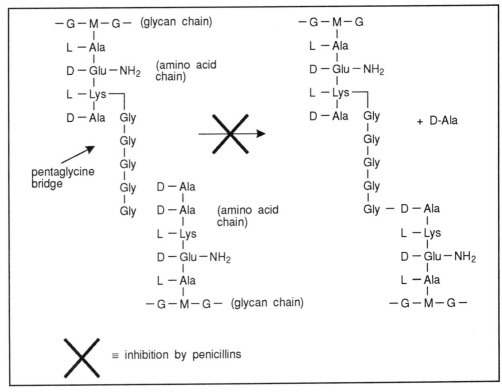

Figure 9.7 Transpeptidation in *Staphylococcus aureus*.

∏   Can you explain why penicillins are bacteriolytic?

Penicillins weaken the cell wall in growing bacteria. Because the cytoplasm of the cell is usually hypertonic in comparison with the extracellular environment, the cell takes up water by osmosis. The weakened cell wall is not able to prevent swelling and the cell bursts (lyses).

Penicillium chrysogenum produces a mixture of penicillins but the growth medium can be altered so that the production of one or two types are vastly enhanced. Examples are shown in Figure 9.8.

Of the penicillin illustrated in Figure 9.8 penicillin G (benzyl penicillin) is a very powerful natural antibiotic, but it has two disadvantages: it is inactivated by the enzyme penicillinase (β-lactamase), and it decomposes in an acid environment (ie the stomach). These problems have been overcome in semisynthetic penicillins. We saw in Section 9.4.2 that the manufacture of improved semisynthetic penicillin was boosted by the isolation of 6-aminopenicillanic acid (6-APA, the basic structure of penicillin) from cultures of *P. chrysogenum*. This compound served as starting material for the development of penicillin derivatives that are resistant to gastric acids and/or penicillinase. Ampicillin has the added advantage of a wide spectrum due to the fact that it can penetrate the outer membrane of the cell wall of Gram-negative bacteria.

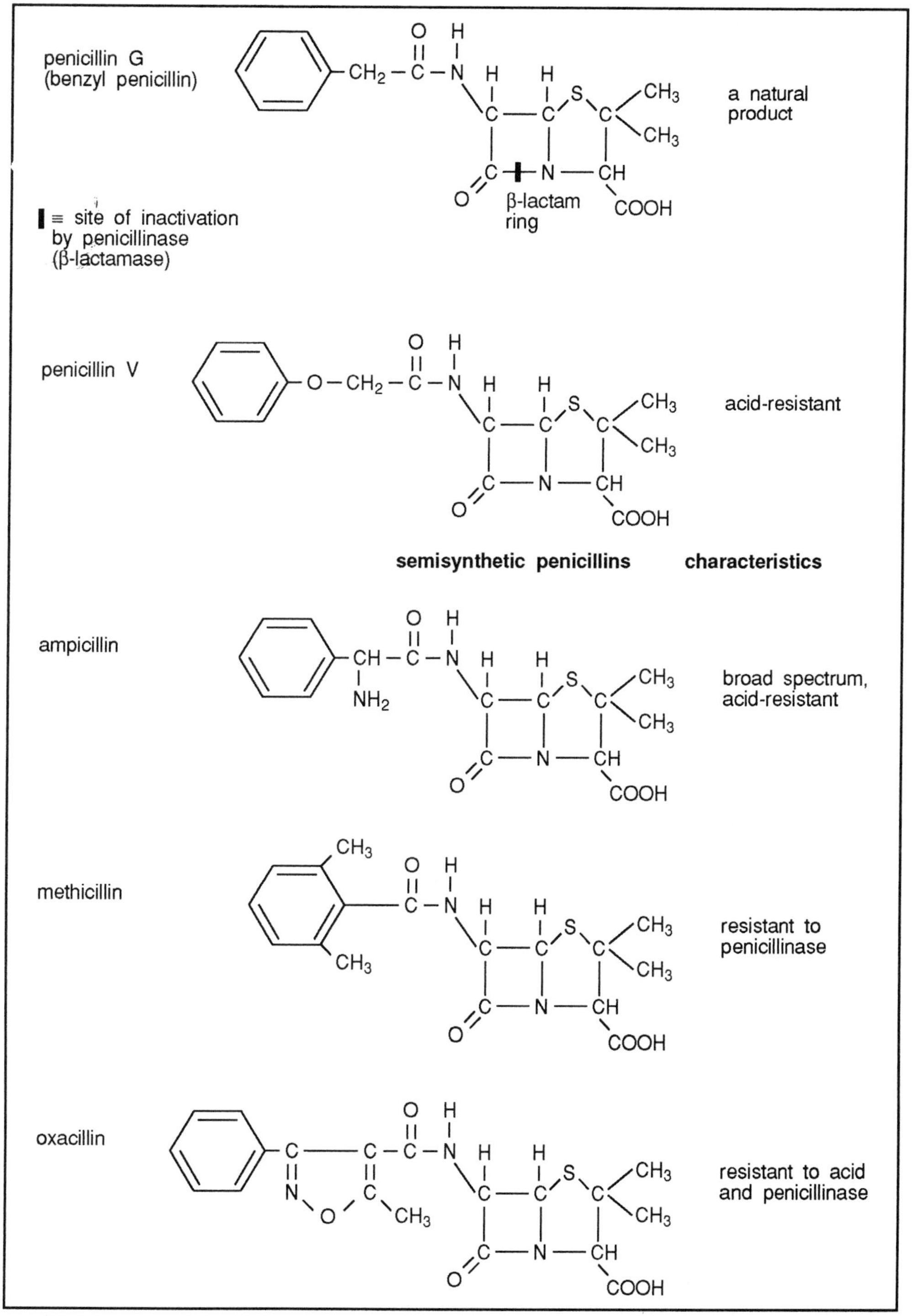

Figure 9.8 Structures of some natural and semisynthetic penicillins (see text for details).

penicillinase
inactivates
penicillin G

Bacteria become resistant to penicillins if they produce penicillinase (β-lactamase), which hydrolyses the β-lactam ring of the penicillin G molecule (Figure 9.8). This enzyme inactivates the antibiotic before it can inhibit cell wall synthesis. Strains of staphylococci produce an extracellular penicillinase, whereas Gram-negative bacteria produce penicillinase in the periplasmic space. These bacteria are penicillin-resistant but can be killed by other antibiotics (for instance, spectinomycin against penicillin-resistant *Neisseria*), or by modified penicillins that are not vulnerable to penicillinases (eg methicillin).

Resistance to penicillin can also be circumvented by combining specific β-lactamase inhibitors with penicillins. For example, clavulanic acid, a β-lactam analogue, produced by a *Streptomyces* inhibits the β-lactamases from several Gram-positive and Gram-negative micro-organisms. It acts by strongly binding to β-lactamase and destruction of the enzyme. Similarly, clavulanic acid is combined with amoxicillin and used as a chemotherapeutic agent against penicillin-resistant bacteria.

Cephalosporins are generally more active and exhibit higher β-lactamase resistance than penicillin but are more toxic and costly to produce.

Antibiotics affecting bacterial cell membranes

The cell membrane in bacteria provides selective permeability of compounds, it is involved in active-transport processes and contributes to maintaining several (vital) ion gradients. Its structure may be disrupted by, for instance, detergents (lipid-dissolving compounds). The so-called ionophores form lipid-soluble complexes with cations so that they affect the ion gradient across the membrane, eg valinomycin renders the membrane permeable to K^+ ions. This type of compound has a great influence on active-transport processes and on energising the membrane.

polymyxins   Polymyxins are cyclic polypeptides, produced by *Bacillus polymyxa* and disrupt the cytoplasmic membrane by binding to phospholipids in the cell membrane, thus interfering with membrane transport. Because of their toxicity they are rarely administered orally, but are mainly applied in ointments, for instance, polymyxin B ointment against skin, ear and eye infections. Although very toxic, they are among the few useful anti fungal antibiotics.

Antibiotics that inhibit bacterial protein synthesis

Protein synthesis in bacteria can be described, simply, as follows. The bacterial ribosome consists of a 50S subunit and a 30S subunit, forming, together with messenger RNA, a 70S ribosome-mRNA complex. Translation starts with the binding of transfer RNA to this complex. After termination of translation the ribosomes revert to their subunit state until they form another complex with mRNA.

Inhibition of protein synthesis by antibiotics usually takes place by binding of the agent to one of the subunits, thus blocking the translation process (Figure 9.6). For instance, streptomycin binds to the 30S unit, whereas erythromycin specifically binds to the 50S unit. Tetracycline prevents the binding of the aminoacyl-tRNAs to the 70S-ribosome mRNA complex. In eukaryotes protein synthesis at the 80S ribosomes can be inhibited by cycloheximide.

Some important antibiotics affecting bacterial protein synthesis will now be considered in more detail:

Example 1: Chloramphenicol

chloram-
phenicol

Chloramphenicol is a naturally occurring broad-spectrum antibiotic, produced by *Streptomyces*. However, it is now chemically synthesised. Chloramphenicol (Figure 9.9) is a bacteriostatic antibiotic which inhibits protein synthesis by binding to the 50S subunit and blocking the reaction catalysed by peptidyl transferase. Chloramphenicol can pass the blood-brain barrier and quickly enter the cerebro-spinal fluid, where it can combat bacterial meningitis.

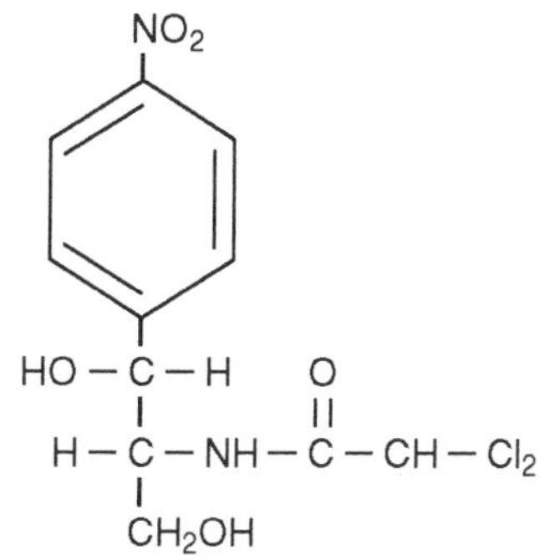

Figure 9.9 Structure formula of chloramphenicol.

Although it is a useful antibiotic, chloramphenicol can be harmful by interfering with the normal development of red blood cells which happens to one in every 50,000 persons.

Example 2: Aminoglycoside antibiotics

streptomycin

Aminoglycosides contain amino sugars linked by glycosidic bonds. Antibiotics in this group include: streptomycin, kanamycin, gentamicin and neomycin. They are used against Gram-negative bacteria, although these days they are not as widely used as when they were first discovered. This is partly a reflection of the development of alterative antibiotics but largely an acknowledgement of their toxicity. In addition the really powerful amingoglycosides such as gentamicin and the semi-synthetic amilcacin are kept 'in reserve' for serious infections.

Streptomycin has also been applied extensively in treating tuberculosis patients but is now, for various reasons (serious side effects, development of resistance), replaced by several synthetic chemicals.

Streptomycin (Figure 9.10) acts by interfering with protein synthesis through binding to a specific protein on the 30S subunit. Resistance against the drug develops when the gene for that protein is mutated in such a way that the streptomycin cannot bind to the ribosome any more. Resistant cells can also produce enzymes (encoded by plasmid-borne genes) which inactivate streptomycin.

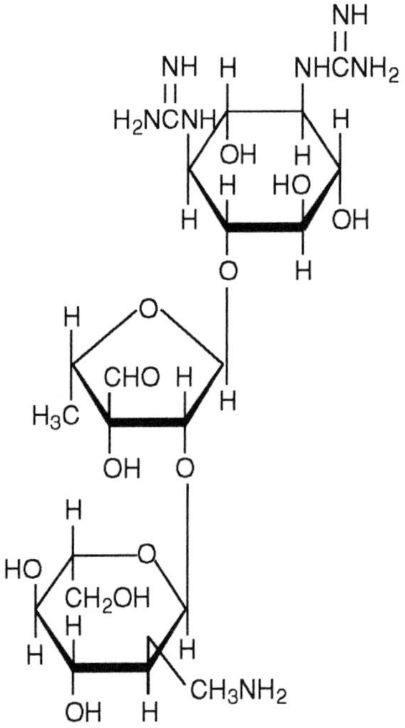

Figure 9.10 Structure formula of streptomycin.

Example 3: Tetracyclines

tetracyclines The basic structure of tetracycline is a naphthacene ring system to which certain chemical groups are added. For instance, compared with tetracycline, chlortetracycline has an H-atom replaced by chlorine (Figure 9.11). These tetracyclines are produced by *Streptomyces* strains. Chemical alteration of the drug enables insertion of other constituents to improve the therapeutic index. The tetracyclines are a group of broad-spectrum antibiotics, inhibiting nearly all Gram-positive and Gram-negative bacteria.

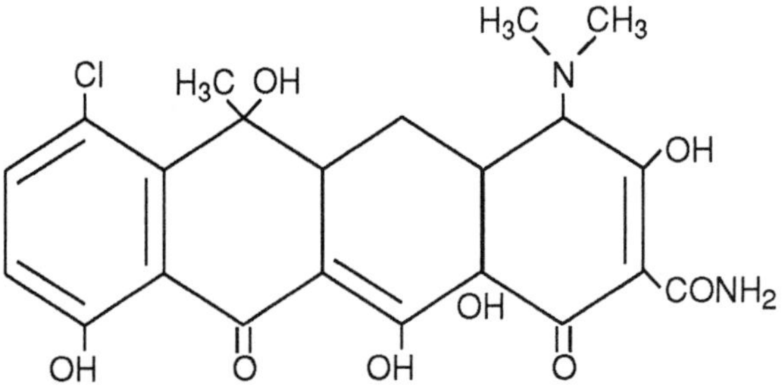

Figure 9.11 The structure of chlortetracycline.

Together with the β-lactams, tetracyclines are the most important group of antibiotics in medicine, despite their low therapeutic index. The tetracyclines' low therapeutic index is due to the fact that they bind with proteins in food and with serum proteins in blood. To overcome this effect relatively high doses need to be given.

The tetracyclines are also used in veterinary medicine, and, in some countries, even as nutritional supplement for poultry and swine.

∏ Can you think why the long-term administration of an antibiotic to prevent the occurrence of a disease should be discouraged?

Such applications carry the danger of developing disease organisms resistant to them.

Another drawback of tetracyclines are their side effects: gastrointestinal discomfort (including diarrhoea), and worse, staining and deformity of developing teeth. Because of the latter effect children under the age of eight and pregnant women are usually not treated with tetracyclines. Overall though we should acknowledge that tetracyclines are relatively non-toxic producing transient or cosmetic problems.

Antibiotics inhibiting nucleic acid synthesis

Actinomycins
mitomycins

Examples of antibiotics inhibiting the synthesis of nucleic acids are actinomycins and mitomycins, which again are produced by strains of *Streptomyces*. Actinomycin D suppresses RNA synthesis whereas mitomycines selectively inhibit DNA synthesis.

SAQ 9.4	Match the antibiotics provided with each of the characteristics labelled 1 to 10: Tetracycline, Penicillin V, Cycloheximide, Streptomycin, Polymixin, Chloramphenicol, Oxacillin, Ampicillin.

1) Broad spectrum.

2) Selective.

3) Bacteriolytic.

4) An acid resistant penicillin derivative.

5) Binds to phospholipids in the cell membrane.

6) Inhibited by β-lactamase.

7) Inhibits eukaryotic protein synthesis.

8) An aminoglycoside antibiotic.

9) Binds to 30S ribosomal subunit.

10) Binds to 50S ribosomal subunit.

9.5 Growth-factor analogues

Some chemosynthetic (produced by chemical synthesis) chemicals are used as chemotherapeutic agents. These are often growth-factor analogues. You will recall from Chapter 2 that growth-factors are organic compounds which the cell cannot produce from simple metabolites but which are essential to growth eg amino acids, purines and pyrimidines, vitamins. Growth-factor analogues are compounds which are very similar in structure to these growth-factors and prevent the utilisation of the growth-factor by acting as a competitive inhibitor. The first growth-factor analogues to be discovered were those of the sulpha group and they proved to be a very successful medicine against various infectious diseases.

Figure 9.12 shows the structure of the simplest of the sulpha drugs, sulphanilamide.

a) sulphanilamide   b) p-aminobenzoic acid

c) folic acid

Figure 9.12 p-Aminobenzoic acid, its analogue sulphanilamide and the growth-factor folic acid.

sulphanilamide inhibits folic acid synthesis

This compound is an analogue of p-aminobenzoic acid, which is part of the vitamin, folic acid. Folic acid is required for biosynthetic reactions by all organisms. If the addition of sulphanilamide generally results in production of the folic acid analogue which results in inhibition of growth why then is this compound not toxic to humans? The answer is that we cannot make our own folic acid and will not absorb the analogue from our diet.

Some examples of other growth-factors and their analogues are given in Figure 9.13.

nucleic acid base analogues

The analogues shown in Figure 9.13 are obtained by adding fluorine or bromine to the growth-factor molecule. The size of the added atom is important for its effect. For instance, the fluorine atom is relatively small so that it does not alter the overall shape of the growth-factor molecule but it does alter its chemical properties. 5-Fluorouracil acts as an analogue of the RNA base uracil and p-fluorophenylanaline as analogue of the amino acid phenylanaline. In contrast, the bromine atom is relatively large (larger

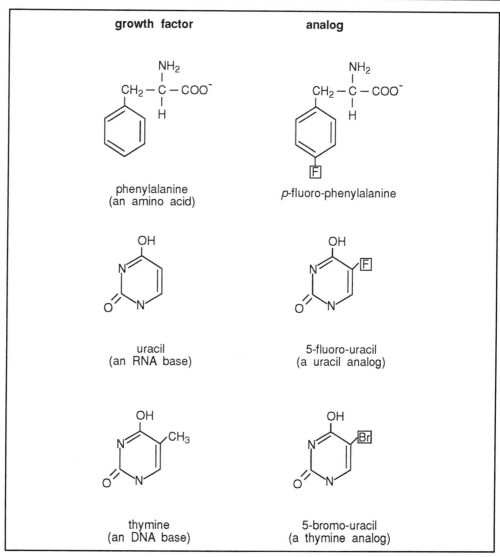

growth factor   analog

phenylalanine
(an amino acid)

p-fluoro-phenylalanine

uracil
(an RNA base)

5-fluoro-uracil
(a uracil analog)

thymine
(an DNA base)

5-bromo-uracil
(a thymine analog)

Figure 9.13 Some growth-factors and their analogues.

than the fluorine atom) and in size resembles the methyl group of thymine - so that 5-bromouracil can act as an analogue of thymine, one of the DNA bases.

∏ Can you think how the action of a growth-factor analogue might be reversed?

The action of growth-factor analogues can often be reversed by adding surplus of the metabolite with structural resemblance. In that case the metabolite can win the competition during biosynthesis.

There are a few analogues of purines and pyrimidines that inhibit replication of certain viruses. For instance, azidothymidine (AZT), is an inhibitor of retroviruses eg the virus causing acquired immunodeficiency syndrome (AIDS). AZT is chemically related to thymidine and after incorporation of AZT in the DNA chain further chain elongation is prevented and multiplication of the virus is inhibited. The use of AZT in treating AIDS is the subject of intensive research and has shown some promise.

∏ Would you expect AZT to have toxic side effects? Give a reason for your response.

Yes. Firstly, because viruses multiply inside host cells (they are obligate intracellular parasites); secondly, host cells would incorporate the thymidine analogue during synthesis of their own DNA.

∏ So, why so you think AZT has any selectivity at all?

The rate of viral DNA replication is faster than of host cell DNA. Therefore, incorporation of the thymidine analogue into viral DNA occurs at a relatively fast rate.

9.6 Distribution and metabolism of chemotherapeutic agents

We already know that the efficacy of a drug is dependant upon many factors relating to drug/micro-organism interaction. In this section we will see that distribution and metabolism of drugs in the body also influences how drugs should be used.

Once administered, the chemotherapeutic agent sooner or later reaches the various compartments of the body, this is schematically shown in Figure 9.14.

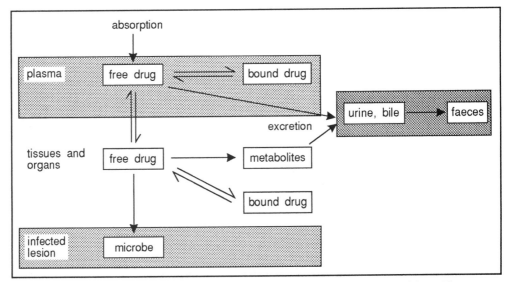

Figure 9.14 Possible pathways of drug through the various compartments of the body. Adapted from Brock, T.D and Madigan, MT 1988, Biology of Micro-organisms, Prentice Hall.

binding to proteins   In the blood a large portion (sometimes up to 90%) is bound reversibly to plasma proteins. Since in a bound state the drug is not active, the minimum inhibitory concentration (MIC) *in vivo* is therefore much higher (sometimes even four or five times) than that in an *in vitro* test system. However, the drug is not irreversibly inactivated by the binding and can be released in active form when the blood level of free drug drops.

detoxification

In the tissue the drug may be metabolised, usually to a less active compound. In fact, detoxified is a better term than metabolised since the body identifies the drug as a foreign agent. The liver is the main detoxifying organ. Enzymes involved in detoxification processes are usually not as well developed in young children as in older patients so that chemotherapeutic drugs are much sooner toxic to the former group. The drug and metabolites derived from it are usually excreted rather rapidly, either by the kidneys and then via the urine, or via the bile (and subsequently via the faeces). Sweat, saliva and milk (in lactating individuals) only play a minor part in this process. As an

drug excretion

example, Figure 9.15 shows the absorption and excretion of penicillin. This pattern is fairly characteristic for drugs. The figure shows that within five hours after administration about 60% of the drug has been excreted in the urine which means that, to maintain an effective level in the body, the drug needs to be administered at regular intervals.

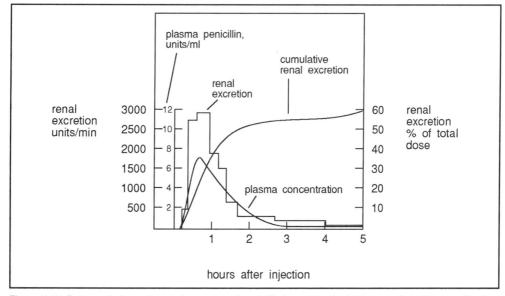

Figure 9.15 Pattern of absorption and excretion of penicillin in a normal adult human. Adapted from Brock, T.D and Madigan, MT 1988, Biology of Micro-organisms, Prentice Hall.

acute and chronic toxicity

Although chemotherapeutic agents are more toxic to the pathogen than to the host, they nearly all have some degree of toxicity for the host. The toxic effect can be either acute or chronic. Acute toxicity shows within a few hours after administration and is, in general, the result of an overdose. Chronic toxicity manifests itself after a long period of time (although sometimes after one or two weeks), as the result of continuous administration. Fortunately, most infectious diseases do not need long-term treatment, with the exception of tuberculosis and certain other chronic (persistent) infectious diseases, which need to be treated for months or even years. High-risk patients may be given chemotherapeutic agents continuously for prophylactic reasons, ie to prevent infections. In these cases, chronic toxicity is often an important aspect of the treatment.

The degree of toxicity of an antibiotic varies greatly between antibiotics. This is partly due to the degree of difference between the microbial cell and the host's cells. Generally antibacterial agents are less toxic than drugs used to combat protozoal (for instance amoebic) or fungal infections where both invading cells and host cells are eukaryotic. Another factor which should be considered is whether the drug evokes an allergic reaction in some patients. This is, for instance, the case for penicillin in 5 to 10% of the human population and influences the therapeutic index of the drug.

in vivo testing of new drugs

Once a new antimicrobial agent has been developed it must be tested for therapeutic activity and toxicity before it can be marketed. Tests are usually carried out on small rodents such as mice. Groups of these animals are infected with a pathogen and the effect of the administered drug monitored. The effect can be measured by simply counting the number of animals that are still alive after various periods of time or by measuring the growth of the pathogen in the animal at intervals during the period of drug administration.

A number of factors need to be taken into account when interpreting the data obtained from such experiments. Firstly, the course of an infection takes in the test animals is usually much faster and more severe than in humans or larger animals since the virulence (ability to cause disease) and the dose of the pathogens used in the testing models are both very high. Secondly, the immunological background of the test animals is usually unknown. However, they are unlikely to have been in contact with the pathogen which means they will not have developed antibodies against it. The value of animal models is that they provide more reproducible and precise quantitative data about the action of a particular chemotherapeutic agent than *in vitro* testing methods.

SAQ 9.5

Identify each of the following statements as True or False. If False give a reason for your response.

1) Sulphanilamide is an analogue of folic acid.

2) 5-Bromouracil is an analogue of the RNA base uracil.

3) High levels of cellular phenylalanine will limit the effect of p-fluorophenylalanine.

4) Binding of an antibiotic to plasma proteins reduces the therapeutic index of a drug.

5) Metabolism of a drug by liver enzymes may increase the therapeutic index of the drug.

6) Oral administration of penicillin G is a suitable route of administration for treatment of *Neisseria gonorrhea* infection in the urinogenital tract.

9.7 Antibiotic resistance

When discussing penicillin (Section 9.4) we mentioned the occurrence of resistance to antibiotics. The resistance can be either natural of acquired. Natural resistance is based on the absence or inaccessibility of the antibiotic's target.

natural and acquired resistance

Acquired resistance develops through a change in the originally vulnerable micro-organism. This type of resistance can develop or be induced in several ways. Firstly, resistance can develop through spontaneous mutation which naturally occurs in most populations. When the antibiotic is administered such mutants are in a favourable position and survive, ie selection takes place. Acquired resistance to streptomycin, for instance, is caused by alterations in one protein in the ribosomal 30S subunit which prevents binding of the antibiotic. Micro-organisms can also be artificially made resistant by treatment with mutagenic agents. Secondly, bacterial resistance against antibiotics may develop by transfer of resistance factors, (R factors or R plasmids). These are extra-chromosomal elements that carry genetic information on antibiotic resistance and that can easily be transferred between related bacteria via conjugation. For instance, the genetic information may concern coding for an antibiotic-inactivating enzyme, such as β-lactamase (which cleaves the β-lactam ring in penicillins and cephalosporins); the action of which was discussed in Section 9.4. The inactivation of antibiotics can also be brought about by covalent modification of the molecule, for instance, by the enzymatic addition of an acetyl group, an adenyl group, or a phosphate group. Kanamycin, gentamicin, streptomycin and other aminoglycosides can be inactivated in this way and the enzymes responsible may also be located on R plasmids. Finally, some microbial cells acquire physical resistance by becoming impermeable to an antibiotic. This kind of resistance is known for erythromycin and tetracyclines.

∏ Can you think how a microbial cell could acquire impermeability to a specific antibiotic?

This could be caused by a mutation leading to loss of function of a transport protein (permease) in the cell membrane, required for uptake of the antibiotic.

misuse of antibiotics

The initially abundant and uncontrolled use of antibiotics once they became widely available, has lead to the development of resistance in many micro-organisms that cause infectious diseases. This is illustrated in Figure 9.16 for the bacterium causing gonorrhoea (*Neisseria gonorrhoea*) which has become resistant for penicillin.

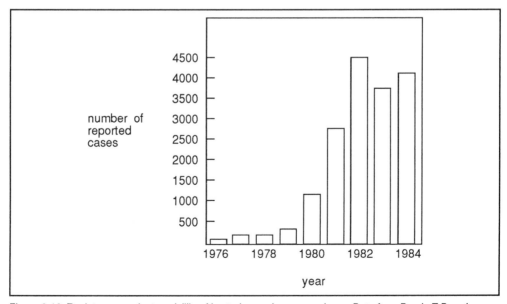

Figure 9.16 Resistance against penicillin of bacteria causing gonorrohoea. Data from Brock, T.D and Madigan, MT 1988, Biology of Micro-organisms, Prentice Hall.

Despite this, antibiotics can still be bought at local grocery shops, without prescription, in many developing countries.

The extensive use of antibiotics in agriculture, as growth-promoting and prophylactic addition to animal feeds, has also contributed to development of drug resistance in bacteria. Some micro-organisms are more apt to develop resistance against antibiotics than others. For instance, staphylococci develop resistance much more readily than streptococci. Bacterial resistance can be minimised by sensible use of antibiotics, ie only for serious infections and at sufficiently high doses to reduce the bacterial population before resistant mutants have a chance to appear. The use of drugs in combinations also prevents the emergence of resistant mutants.

Π Resistant mutants occur at a frequency of 1 in 10^6 for drug A and of 1 in 10^8 for drug B. If these two drugs were used in combination, at what frequency would you expect resistant to both drugs?

Resistant mutants are likely to occur a frequency of 1 in 10^{14} ($1 \times 10^8 \times 1 \times 10^6$).

the search for new antibiotics

Bacterial resistance is the main reason why the search for new antibiotics still goes on. One such new antibiotic is aztreonam, a member of a new family of antibiotics called monolactams. During an extensive search, a bacterium was isolated from a river in New Jersey which produced a new compound containing a β-lactam ring. Although the compound was very unstable it proved to be an important discovery because it was resistant to β-lactamases. Chemists managed to stabilise the compound by chemically converting it and the new antibiotic aztreonam was born. It is an effective antibiotic, especially against aerobic, Gram-negative bacteria and has a high therapeutic index. It has the added advantage of not causing allergic reactions in people who are allergic to penicillin.

SAQ 9.6

Identify each of the following statements as True or False. If False give a reason for your response.

1) Eukaryotic cells are naturally resistant to tetracycline because their protein synthesis mainly involves 80S ribosomes.

2) Prophylactic treatment using methicillin will lead to the rapid emergence of resistant strains.

3) R factors transfer antibiotic resistance from Gram-positive to Gram-negative bacteria and *visa versa*.

4) Bacterial endospores have an acquired resistance to certain disinfectants.

Summary and objectives

In this chapter we have learnt that there are many different kinds of antimicrobial agents. Disinfectants are generally non-selective and can only be applied to inanimate objects. The chemical properties of the various classes of disinfectants largely governs their use. Conversely, chemotherapeutic agents are selective and are used for the treatment of infections. The selective toxicity of these drugs is determined by their mode of action which generally exploits a structural or process difference between mammalian host cells and the infectious agent. Some drugs are synthesised naturally by micro-organisms but many are now produced partially or wholly by chemical synthesis. The use of drugs is governed by a complex interaction between drug/host/infectious agent and is often a compromise between desired antimicrobial effect and degree of toxicity to the host. In the case of the antibiotics, microbes may acquire resistance through spontaneous mutation or plasmid transfer leading to increased inactivation of the drug.

Now that you have completed this chapter you should be able to:

- explain the difference between the various antimicrobial agents (antiseptics, disinfectants, antibiotics and chemotherapeutic agents);

- differentiate between the effects these agents have on sensitive organisms (static, cidal and lytic effects);

- describe the mode of action of the various antimicrobial agents and give examples of agents for each mode of action;

- describe the survival kinetics of micro-organisms mathematically;

- give a description of the methods used to measure antimicrobial activity and derive values from supplied data;

- list factors influencing the effect of antimicrobial agents;

- list characteristics of common disinfectants, antiseptics and chemotherapeutic agents;

- define the following terms: broad-spectrum and narrow-spectrum antibiotic, decimal reduction time, natural and acquired resistance, phenol-coefficient, therapeutic index;

- explain how acquired resistance to antibiotics can develop.

Responses to SAQs

Responses to Chapter 1 SAQs

1.1

Approximate date	Researcher	Nature and significance of the work
1540	Fracastoro	Suggested that disease is caused by living organisms
1650	van Leeuwenhoek	First accurate recordings of observations of micro-organisms using a microscopy
1680	Redi	Proved experimentally that maggots arise form flies' eggs and not by spontaneous generation
1825	Bassi	Showed that a fungus was the cause of a silkworm disease
1837	Scwann	Proposed that yeast cells convert glucose to ethanol
1851	Pasteur	Proved that yeast cells convert glucose to ethanol
1860	Lister	First use of an antimicrobial agent (phenol) on wound dressings and in operating theatre. Heat sterilised instruments
1861	Pasteur	Proved that microbial spoilage was caused by micro-organisms from air contamination and not arising by spontaneous generation
1876	Koch	Proved *Bacillus anthracis* caused anthrax. Published Koch's postulates. Introduced the concept of pure culture. Introduced the concept of solid (agar or gelatin) medium
1875-1885	Pasteur	Developed rabies and anthrax vaccines
1884	Cumberland	Introduced the porcelain bacterial filter
1887	Petri	Introduced the petri dish

1890 onward	Winogradsky	Pioneering work in chemoautotrophs
1890 onward	Beijerinck	Pioneering work on *Rhisobium* and *Azotobacter spp.* With Winogradsky developed the enrichment technique
1897	Buchner	First demonstration of cell free extracts carrying out a multi-stage chemical process
1900	Beijerinck	Published definitive work on tobacco mosaic virus
1900	Loeffler	Published definitive work on foot and mouth virus
1917	d'Herelle	Coined the term bacteriophage for viruses which infect bacteria
1935	Stanley	Crystallised the TMV virus showing it contained only protein and RNA.

1.2   The simple answer to all seven parts of SAQ 1.2 is true! Looking at one or two of these statements it may seem hard to believe that they are true and in some cases there are only a very limited number of bacteria which are applicable.

1) True. Most compounds are relatively rapidly degraded although a few compounds may be difficult to degrade and the degradative process may be relatively slow and require specific conditions. Notable amongst organisms know for their versatility are members of the family Pseudomonadaceae.

2) True. The extremes at which microbes will not only exist but also grow and divide mean that virtually all habitats on Earth are suitable for at least some organisms.

3) True. Bacteria have been found which grow in chimneys or (fissures located deep on the ocean floor). These chimneys expel sulphide at temperatures over 300°C and pressures of up to 250 atmospheres. Bacteria growing very close to these chimneys grow at 240°C, 250 atmospheres pressure with a generation time of only 40 minutes!

4) True. Examples of bacteria (*Thiobacillus*), fungi (*Aconitum*), and algae (*Cyandium*) are known which grow in acid hot springs at pH values below 1.0.

5) True. Experiments have been carried out which demonstrate that virtually all bacteria are relatively unaffected by very high pressures. Only those organisms which do not have a cell wall, for example most of the protozoa, cannot withstand such pressures.

6) True. Many of the tasks which involve recycling of elements through various organic/inorganic or different oxidation phases are only carried out by micro-organisms, particularly bacteria and fungi.

7) True. Many bacteria, particularly *Pseudomonas spp.*, can divide in less than 25 minutes with the fastest thought to divide in about 17 minutes. The following example will be described in more detail in Chapter 4. If a single cell (weighing 10^{-13}g) and its progeny divide every 20 minutes, after 44 hours there will have been 44 x 3 or 132 generations (that is divisions of the original cell and its ancestors). As 1 cell divides into 2 cells and they each divide to give 4 cells etc the number of cells after 132 generations is 2^{132} which when calculated and corrected for the weight of an average cell gives a value equivalent to the weight of the Earth.

Responses to Chapter 2 SAQs

2.1   First of all see completed Table 2.1 below.

Elements contained	Monomers	Polymers	Percentage of the dry weight
carbon + hydrogen + oxygen + nitrogen + some sulphur	amino acids	proteins	55%
carbon + hydrogen + oxygen + nitrogen + phosphorus	nitrogenous bases - purines and pyrimidines	nucleic acids	23%
carbon + hydrogen + oxygen (+ some phosphorus)	fatty acids	lipid + phospholipid	9%
carbon + hydrogen + oxygen	monosaccharides	polysaccharides (carbohydrates)	5%

1) The almost infinite number of proteins are composed, virtually without exception, of mixtures of the twenty one amino acids common to all living systems.

2) Nucleic acids would be either deoxyribonucleic acid (DNA) or ribonucleic acid (RNA) and typically the dry weight of RNA exceeds that of DNA by a ratio of around 6:1.

Proteins act as both structural entities, being a component of membranes; and also as the catalysts or enzymes of the cell.

The deoxyribonucleic acid within cells is the carrier of the hereditary information of the cell. The ribonucleic acid exists in one of three forms; messenger (m-RNA), ribosomal (r-RNA) and transfer (t-RNA). These three are involved in transporting and translating into protein the information provided by the DNA.

Lipids, principally as phospholipid, are found in the membranes of all living cells. Neutral lipids in higher organisms, though not generally in bacteria, also serve as important energy storage compounds.

Polysaccharides or carbohydrates serve as storage materials (for example glycogen and starch) in many organisms and may also serve as structural entities, for example the peptidoglycan polymers in bacterial cell walls.

2.2   1) The four categories are: photoautotrophs, photoheterotrophs, chemoautotrophs, chemoheterotrophs.

2) A photoautotroph is an organism which utilises inorganic compounds as principal carbon source and light as an energy source.

A photoheterotroph is an organism which utilises light as an energy source and organic compounds as principal carbon source.

A chemoautotroph is an organism which utilises chemicals as an energy source and inorganic compounds as principal carbon source.

A chemoheterotroph is an organism which utilises chemicals as an energy source organic compounds as principal carbon source.

Note that the two sub-divisions of chemotrophs distinguish between organic and inorganic sources of carbon but not between organic or inorganic sources of energy.

3) Photoautotrophs include all of the higher plants, eukaryotic algae, the prokaryotic blue-green algae (cyanobacteria), together with some photosynthetic bacteria - the purple sulphur and green sulphur bacteria. Overall a very large group of organisms all of which use carbon dioxide as principal carbon source and light as energy source.

Photoheterotrophs are a restricted group containing only the purple non-sulphur photosynthetic bacteria and some members of the green sulphur bacteria. By definition they use organic carbon sources though some can use carbon dioxide as a partial or total substitute.

Chemoautotrophs are restricted to a few genera of very common bacteria which are important in nature. These include the nitrifying bacteria, iron bacteria, sulphur-oxidising bacteria and hydrogen-oxidising bacteria.

Chemoheterotrophs are the largest of the groups containing all animals, protozoa, fungi and the overwhelming majority of bacteria.

2.3

1) Ammonium chloride is added to each of the media A, B and C as a source of nitrogen. Although it provides additional hydrogen and chlorine, these are more than adequately catered for in the common ingredients.

Glucose is added as a carbon and/or energy source. The hydrogen and oxygen will also be utilised but have already been provided by alternative means.

Nicotinic acid is only required in small amounts. It is an example of a growth factor, being a precursor of the enzyme cofactor nicotinamide adenine dinucleotide or NAD^+.

Yeast extract will be discussed in detail later but is essentially a complex mixture of nutrients including amino acids, vitamins, macro-nutrients and trace elements.

2) There is still no evidence of the presence of carbon in medium A. Organisms obviously require carbon for growth and the only source will be carbon dioxide. Thus no organic carbon source will be available for organisms growing on medium A.

2.4

1) Medium A contains no apparent carbon source and therefore only atmospheric carbon dioxide is available as a carbon source. Organisms growing in this

environment would be chemoautotrophic - for example nitrifying bacteria such as *Nitrosomonas* which obtain energy from the aerobic oxidation of ammonia.

This is the most difficult medium to find an answer for; but you should always try to follow the same stages; note if all necessary elements are present, identify the carbon source, energy source and see whether conditions are aerobic or anaerobic.

Medium B is far more versatile in that it contains glucose as a carbon and energy source. Thus chemoheterotrophic organisms - many bacteria and protozoa and fungi would grow on this medium provided that they did not require growth factors. Of course chemoautotrophic organisms will also grow on this medium.

Medium C is similar to B but allows growth of chemoheterotrophic organisms which require nicotinic acid.

Medium D contains yeast extract which supplies many growth factors and is thus a very versatile medium for the growth of chemoheterotrophs.

2) The absence of molecular oxygen as an electron acceptor renders medium A unable to support growth of organisms which obtain their energy by aerobic respiration. However media B, C and D will support growth of anaerobic or facultative organisms which ferment glucose rather then oxidise it to yield energy.

2.5 A general purpose medium is one which contains sufficient nutrients to support the growth of many micro-organisms without inhibiting or specifically encouraging any particular organisms.

Enrichment or enriched media are those in which either a specific macro-nutrient has been added to encourage growth of particular organisms or the environmental conditions have been modified. The emphasis is on encouraging certain types rather than inhibiting others.

Selective media are those which are selective in that they usually contain one or more chemicals to suppress certain organisms whilst allowing the growth of others. Enrichment media are in fact a type of 'passively selective' medium.

Differential media are those which distinguish between different groups of organisms and may even allow tentative identification. Such media are often selective as well.

Diagnostic media are those specifically designed to investigate one or a few properties of a given micro-organism. For instance if we are trying to distinguish between two closely related organisms, we can carry out a series of biochemical tests - each possibly requiring its own diagnostic medium - to encourage growth and subsequent elucidation of a specific positive or negative result.

2.6

Type of organism	Examples of micro-organisms	Natural habitat
psychrophile	Some bacilli, *Pseudomonas spp.*, other bacteria; some yeasts.	Polar regions, oceans, lakes, cold soils. Commercially a nuisance in refrigerated environments (eg as food contaminants).
mesophile	The majority of micro-organisms including parasites and pathogens of higher animals and plants.	Generally in temperate and hot climates and associated with man and his environment. Includes the organisms of medical importance.
thermophile	Bacteria in particular including some of the spore formers - from *Bacillus* and *Clostridium spp.*	Hot springs and natural habitats where the temperature regularly (but not necessarily constantly) gets above 50°C. In man made heating or heated systems.

2.7 Obligate aerobes include virtually all eukaryotes except those listed below, together with many bacteria.

Facultative anaerobes include many bacteria, for example members of the Enterobacteriaceae. The only common eukaryotes in this group are yeasts of the genus *Saccharomyces* which will oxidise glucose to CO_2 or, in the absence of molecular oxygen, ferment glucose to ethanol and CO_2.

Obligate anaerobes include several genera of bacteria, for example the genus *Clostridium*. Obligately anaerobic eukaryotes are very rare and principally include a few 'rumen protozoa' - organisms living in the rumen of higher animals such as cows and sheep.

Aerotolerant organisms are a group which may be confused with aerobes because their natural environment is often aerobic. Examples here are members of the lactic acid bacteria, for example *Lactobacillus spp.*, *Leuconostoc spp.* and *Streptococcus spp.* all of which ferment glucose to a variety of end products.

Micro-aerophilic organisms include the well-studied *Rhizobium spp.*, bacteria which require oxygen for aerobic respiration and are able to fix atmospheric nitrogen. However their nitrogenase enzyme is inactivated by modest concentrations of molecular oxygen. In the laboratory their environment has to be strictly regulated, in nature they are found inside higher plant root nodules in a suitable low concentration of oxygen.

2.8 The most important factor regarding the choice of medium is that the organism must be able to fix atmospheric nitrogen. Thus the medium will not contain a fixed nitrogen source, however all other nutrients must be provided.

The most likely habitat is the soil if we require nitrogen fixing organisms. Generally one would choose a well-ventilated, fertile soil as this will contain more viable organisms in total than barren soils. A small sample of soil therefore would either be shaken in a broth culture or spread over an agar plate.

The choice of incubation conditions will be influenced by the source of the culture, in this case the soil. As organisms fix atmospheric nitrogen, they would normally be exposed to air and therefore they should be incubated aerobically. Organisms which occur predominantly in soil are psychrophilic and thus the incubation temperature should be 30°C or less and optimally around 22°C. Soil pH can vary but fertile soils are generally not far from neutral. Thus a medium of neutral pH, incubated aerobically at 22°C would be indicated.

2.9   *Bacillus* species are soil organisms having few if any unusual nutritional requirements. Being heterotrophs they require an organic carbon and energy source, ammonium ions and a balanced mineral medium. Coming from the soil, we would anticipate incubation at 22 - 30°C though a few bacilli are thermophilic. However many organisms would grow under the conditions described above without giving us any apparent enrichment.

The key to isolating bacilli is connected with the fact that their spores are heat resistant. This can be exploited by pre-heating the inoculum for a short time in order to kill off ordinary or vegetative cells but leaving spores as survivors. In this case heating the inoculum to about 80°C for 2 to 5 minutes is suitable. On cooling and inoculating the media, the spores of the one or more species of bacilli in the inoculum will then germinate.

Responses to Chapter 3 SAQs

3.1   Before incubation:

$$\frac{126 + 93 + 81}{3} \times 10^6 \times 10 = 1.0 \times 10^9 \text{ CFUs ml}^{-1} \text{ (x } 10^6 \text{ is the dilution, x 10 because we used}$$
0.1ml of sample)

After incubation:

$$\frac{102 + 91 + 107}{3} \times 10^8 \times 10 = 1.0 \times 10^{11} \text{ CFUs ml}^{-1}$$

3.2   1) MPN.

2) Haemocytometer or slide culture.

3) Coulter counter.

4) Pour plate, spread plate and MPN.

5) Pour plate, spread plate, MPN and slide culture.

6) Slide culture.

7) Pour plate.

8) Coulter counter (and possible haemocytometer).

9) All methods listed.

10) MPN.

3.3 1) The plot of absorbance against dry weight is obtained as follows.

Dry weight of cells in undiluted culture = (2.8456 - 2.7656)/10 = 8 mg ml^{-1}. Dry weight values are obtained from this concentration by taking into account the dilution.

The absorbance of medium minus cells is subtracted from each of the other absorbances. These values are then plotted against the corresponding dry weight values.

Thus your data should be:

culture density (mg ml^{-1})	corrected absorbance (at 540nm)
0.8	0.990 (from 1.05 - 0.06)
0.57	0.910
0.4	0.770
0.32	0.660
0.27	0.535
0.2	0.410
0.114	0.230

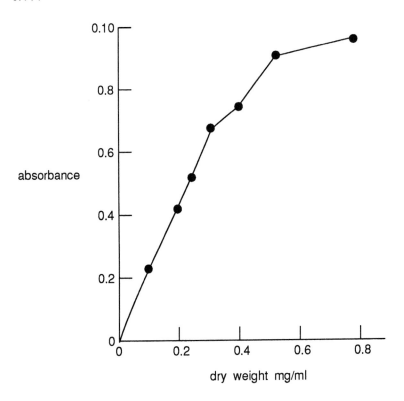

2) Absorbance due to cells = 0.560 - 0.060 = 0.50. We can see from the graph that an absorbance of 0.5 is equivalent to a dry weight of 0.24 mg ml^{-1}. Since the culture was diluted five-fold, the culture density is 0.24 x 5 = 1.2 mg ml^{-1}.

3.4

1) Protein: complex growth medium.

2) ATP: slow sampling speed.

3) Absorbance: cell clumping; the presence of non-cellular particulate material; cell lysis; changes in biomass morphology.

4) Wet weight: changes in the morphology of the cells; cell lysis; the presence of non-cellular particulate material.

5) DNA: none.

3.5

50 mg: dry weight

1.0 mg: Biuret protein and DNA

0.1 mg: Lowry protein and absorbance

0.00001 mg: cell count

3.6

1) False. An estimation of dry weight can be obtained by dividing PCV measurements by five.

2) True.

3) True.

4) False. Wet weight measurements are usually less accurate than dry weight measurements because extracellular water content is difficult to standardise.

5) True. Perhaps an explanation is required. The statement is true because the sensitivity of measurement increases although the accuracy for cell mass determintaion may decrease due to increased interference from cell components.

6) False. The sampling time should be less than 1 second to minimise error caused by rapid turnover of the ATP pool. We pointed out for example that the cell's pool of ATP may be turned over 4-8 times per second!

3.7

1) Poor accuracy of measurement of nutrient or product.

2) High growth yield coefficient.

3) Variable growth yield coefficient.

4) Substrate also used for product synthesis.

5) Chemically unstable substrate or product.

6) Metabolism by resting cells ie substrate consumption or product formation poorly correlated with growth.

7) Product further metabolised ie not an end-product of metabolism.

8) Growth in complex media in which more than one component serves as carbon source.

9) Some substrate converted to cell storage material.

3.8 We use the relationships:

Biomass concentration $(X) = \dfrac{OTR}{\mu} \cdot \dfrac{Y_o}{k}$ (Equation 1)

where:

k = correction factor = 0.0313 mol O_2 g^{-1} O_2

μ = specific growth rate = $0.2h^{-1}$ (since μ = dilution rate in continuous culture)

Oxygen transfer rate is determined using:

$OTR = Q\,(C_{in} - C_{out})$ (Equation 2)

$OTR = 1 \;\cdot\; 0.5 = 0.5$ mmol O_2 min^{-1}
$\qquad\qquad = 30$ mmol O_2 h^{-1}

Y_o is determined using:

$$\frac{1}{Y_o} = 16\left[\frac{2\,C + H/2 - O}{Ys \cdot M} + \frac{O'}{1600} - \frac{C'}{600} + \frac{N'}{933} - \frac{H'}{200}\right]$$ (Equation 3)

$$\frac{1}{Y_o} = 16\left[\frac{2 \cdot 6 + 12/2 - 6}{0.5 \cdot 180} + \frac{20}{1600} - \frac{50}{600} + \frac{14}{933} - \frac{8}{200}\right]$$

$$\frac{1}{Y_o} = 16\,(0.0375)$$

$Y_o = 1.67$ g biomass $g^{-1}O_2$.

Substituting into equation 1), we have:

$$X = \frac{30}{0.2} \cdot \frac{1.66}{31.3} = 7.9\ g$$

3.9

1) Measurement of O_2 consumption has merits a), b), c), d) and, in some cases, e) but has limitations ii).

2) Dry weight has none of the merits listed but has limitations i), iv), v).

3) Protein measurement has merit b) but limitations i), iii).

4) Slide culture has merits b), c), d) but limitation v).

5) ATP measurement has merits b), c) but limitations i), ii).

Responses to Chapter 4 SAQs

4.1

1) a) $n = \dfrac{\log N_t - \log N_o}{\log 2} = \dfrac{9 - 2}{0.3010} = 23.2$

b) $t_g = \dfrac{t}{n} = \dfrac{10}{23.2} = 0.43 \text{ hours}$

c) $k = \dfrac{1}{t_g} = 2.32 \text{ h}^{-1}$

d) $\mu = 0.693\,k = 1.61 \text{ h}^{-1}$
     or $\mu = 0.693/t_g = 1.61 \text{ h}^{-1}$

e) $\dfrac{dx}{dt} = \mu X$ thus $\dfrac{dx}{dt} = 1.61 \times 10^9 \text{ cells h}^{-1}$

For determinations of t_g, k, μ and growth rate to be valid, we must assume that the culture is growing exponentially throughout the incubation period.

2) $n = \dfrac{\log N_t - \log N_o}{\log 2} = \dfrac{6 - 2}{0.3010} = 13.3$

Since $t_g = \dfrac{t}{n}$ then $t = t_g.n$

$t = 15 \;.\; 13.3 = 199$ minutes.

4.2   1)

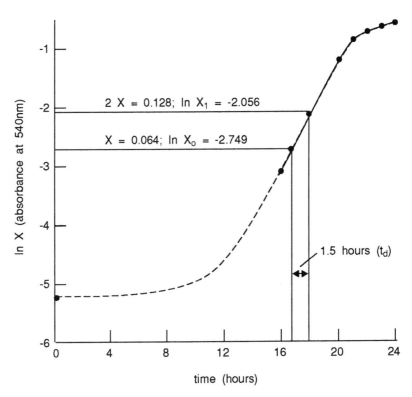

time (hours)

2) The doubling time (t_d) can be read from the graph: 1.5 hours

$$t_d = \frac{\ln 2}{\mu} \text{ so } \mu = \frac{\ln 2}{t_d} = \frac{0.693}{1.5} = 0.46\ h^{-1}$$

3) We an see from the graph that 11 hours would be a reasonable estimate of the length of the lag phase.

4.3   1) False. The growth rate increased during exponential phase because the biomass concentration is increasing:

$$\frac{dx}{d_t} = \mu X$$

However, the specific growth rate (μ) is constant during exponential growth.

2) True.

3) False. See Section 4.3.3. For cyanobacteria and algae some cells are dividing whilst others are dying. This gives rise to a constant viable cell number during stationary phase but a gradual increase in total cell number. So, for these cultures, biomass concentration (total cell number) increases with time in the stationary phase.

ie ($\frac{dx}{dt} > 0$) for these cultures.

4) True.

5) True.

6) False. K_s determines the specific growth rate at low (non-saturating) nutrient concentrations.

4.4   1) The growth yield coefficient can be determined from a plot of glucose concentration versus biomass concentration, or from $Y_s = \dfrac{\text{biomass produced}}{\text{substrate used}}$.

Slope = Y_s = 0.5 g biomass (g^{-1} glucose)

2) Using: $q = \dfrac{\mu}{Y_s}$

we obtain, $\mu = q . Y_s$,
$\mu = 2.5 . 0.5 = 1.25$ h^{-1}

4.5

	Methanol	Ethanol	Methane
Molecular mass	32	46	16
Y_{molar} (g mole^{-1})	0.4 x 32 = 12.8	0.3 x 46 = 13.8	0.8 x 16 = 12.8
Carbon in substrate (%)	37.5	52.2	75.0
Carbon conversion efficiency (%)	$\left(\dfrac{45}{37.5}\ 0.4\right)$ x 100 = 48	$\left(\dfrac{45}{52.2}\ 0.3\right)$ x 100 = 26	$\left(\dfrac{45}{75}\ 0.8\right)$ x 100 = 48

4.6   1) a) When μ equals 0.1 h^{-1} the generation time is 6.93 hours (using $t_g = 0.693/\mu$). Since the inoculum consisted of exponentially growing cells derived from the same medium one would not expect a lag phase. This means the culture has produced 50/6.93, or over 7 generations (7.215).

The weight of inoculum (X_o) can be calculated using equation 4.9.

$$\mu = \frac{\ln X_t - \ln X_o}{t}$$

Rearranging we have:

$\ln X_o = \ln X_t - \mu t$. Since $X_t = 2$g^{-1}, t = 50h and $\mu = 0.1$h^{-1}

$\ln X_o = \ln 2 - 0.1 . 50$

$$= 0.693 - 5 = -4.307 \text{g}^{-1}$$

$$X^\circ = e^{-4.307} = 13.4 \text{ mg}^{-1}$$

b) $Y_{molar} = \dfrac{X_t - X_o}{S_t - S_o}. \text{ M glucose}$

$$\dfrac{2 - 0.0134}{20 - 0.25}. \; 180 = 18.1 \text{ g (mole substrate)}^{-1}$$

c) $Y_{ATP} = \dfrac{Y_{molar}}{\text{ATP per mole substrate}}$

$$= \dfrac{18.1}{2} = 9.0 \text{ g biomass mole}^{-1} \text{ ATP}$$

2) Since $Y_{ATP} = \dfrac{Y_{molar}}{\text{ATP per mole substrate}}$. Then ATP per mole substrate $= \dfrac{Y_{molar}}{Y_{ATP}}$

$$\dfrac{28}{9.3} = 3 \text{ moles ATP mole}^{-1} \text{ substrate}$$

4.7

1) True (Section 4.6.1)

2) True. (Figure 4.7)

3) True. (Figure 4.7)

4) True. (Figure 4.7)

5) False. At high growth rates daughter cells may receive more than one copy.

6) True (Section 4.6.1)

7) False. (Section 4.6.3). Mean cell cycle time tells us nothing about the degree of population synchrony. It is merely the average of the generation times of all of the individual members of the population.

Responses to Chapter 5 SAQs

5.1

1) For A $D_c = 0.495 \text{h}^{-1}$ and for B $D_c = 0.262 \text{h}^{-1}$. The critical dilution rate D_c can be calculated using Equation 5.10 ($D_c = \mu_{max} \dfrac{S_R}{(K_s \, S_R)}$. The glucose concentration in the reservoir (0.2%) equals 2 gram glucose per litre medium, $S_R = 1.1.10\text{-}2 \text{ mol l}^{-1}$. At K_s $10^{-4} \text{ mol l}^{-1}$, this gives us $D_c = 0.495 \text{ h}^{-1}$; at $K_s = 10^{-2} \text{ l}^{-1}$, $D_c = 0.262 \text{ h}^{-1}$.

2) a) In steady state $\mu = D$.

$$\mu = D = \dfrac{F}{V} = \dfrac{1}{2} = 0.5 \text{ h}^{-1}.$$

b) Using equation 5.5:

$$\bar{s} = K_s \cdot \frac{D}{\mu_{max} - D}$$

$$0.05 = K_s \frac{0.5}{2 - 0.5}$$

$$K_s = 0.15 \text{ g l}^{-1}$$

c) Using equation 5.8:

$$\bar{x} = Y (S_R - \bar{s})$$

$$2 = Y (5 - 0.05)$$

$$Y = \frac{2}{4.95} = 0.4 \text{ g biomass g}^{-1} \text{ substrate}$$

5.2   The stages are:

1) operate the chemostat in steady-state at different dilution rates and measure biomass concentrations $(\bar{x})$ and substrate concentrations $(\bar{s})$;

2) use equation 5.8 to generate a value for Y at each dilution rates (S_R value must be known);

3) plot $\dfrac{1}{Y}$ against $\dfrac{1}{D}$;

4) slope of line = m
   Intercept with Y ordinate $= \dfrac{1}{Y_G}$;

5.3   1) False. At dilution rates approaching D_c biomass productivity $(D\bar{x})$ declines sharply. (see Figure 5.6).

2) False. Slope $= K_s/\mu_{max}$ intercept on the $\dfrac{1}{s}$ axis $= -\dfrac{1}{K_s}$ (see Section 5.4).

3) True.

4) False. $\bar{s}$ increases sharply as μ approaches μ_{max} (see Figure 5.6).

5) False. Increasing S_R concentration will increase the biomass concentration $(\bar{x})$ in the vessel. Specific growth rate is increased by increasing the dilution rate.

6) False. However, the specific growth rate equals the dilution rate.

7) False. Maintenance coefficient influences biomass yield most markedly at low dilution rate (see Figure 5.8).

8) True.

9) True.

10) True. Since turbidostats are generally operated at higher dilution rates than chemostats (see Figure 5.6).

11) False. $D = \dfrac{F}{V}$ for a chemostat. In a fed-batch culture $D = \dfrac{F}{V_o + F \cdot t}$.

12) False. The maintenance coefficient may be dependent upon growth rate, in which case a non-linear $\dfrac{1}{Y}$ against $\dfrac{1}{D}$ plot is obtained.

5.4 A constant low nutrient level is favourable for 2, 3 and 4. It is unfavourable for 1 since micro-organisms would grow at submaximal growth rates.

5.5
1) The organism with the highest specific growth rate at a certain $\bar{s}$ will become dominant. μ can be calculated using Equation 5.3, which gives the following values:

$\bar{s}$ (mol l^{-1})	μ (h^{-1})		
	A	B	C
0.33×10^{-4}	0.021	0.016	0.014
1.4×10^{-3}	0.131	0.274	0.259
0.45×10^{-2}	0.144	0.375	0.366

From these data it is clear that at the lowest steady-state substrate concentration micro-organism A wins the competition. In both other cases B has the highest growth rate.

2) It is impossible to make organism C dominant since μ_B will in all cases be higher than μ_C. Also, increase in dilution rate does not lead to dominance of C since D_c at a certain value for S_R for organism B is always greater than that for C (see Equation 5.4).

5.6
1) $\mu_{max} = 1.0$ h^{-1} ie D_c without biomass feedback.

2) Concentration factor $= \dfrac{\bar{x}_2}{\bar{x}_1} = \dfrac{1.0}{0.5} = 2.$

3) Using $\mu = A\,D$

Concentration factor $= \dfrac{1}{A}$

So, $A = \dfrac{1}{2} = 0.5.$

If $D = 1.5$ then $\mu = 0.5 \cdot 1.5 = 0.75$ h^{-1}.

4) Using $\mu = [c(1-h)+h]\,D$

Assuming $\mu = 0.75\ h^{-1}$, then $D = 1.5h^{-1}$, $h = 0.2$ (remember that if $h = 1$ there is no feedback, so 80% filter efficiency is equivalent to $h = 0.2$).

c = fraction of outflow not filtered.

$0.75 = [c(1-0.2)+0.2]\,1.5$

$0.75 = c.1.2 + 0.3$

$$c = \frac{0.75 - 0.3}{1.2} = 0.38$$

5) We can see from Figure 5.14 that maximal productivity is obtained at a dilution rate of around $1.8h^{-1}$. Productivity (output rate) $= D\bar{x}_2$ or $D\bar{x}_1\,A\left(as\ \dfrac{1}{A} = \dfrac{\bar{x}_2}{x_1}\right) = 1.8 \times 1 = 1.8g$ biomass h^{-1}.

5.7

1) System b and e (for second stage).

2) System d (for second stage).

3) System b and e (for second stage).

5.8

1) A single-stream multi-stage system is appropriate, since low or zero growth can be achieved in the second stage.

But, in practice, many antibiotic fermentations are operated as repeated fed-batch cultures. These cultures have phases when the growth rate is close to zero. They are also generally more stable than multi-phase systems.

2) A simple chemostat is appropriate, since the growth rate can be altered by merely changing the dilution rate.

3) Fed-batch culture may be appropriate, since the limiting nutrient is maintained at a very low level. Alternatively, a single-stream multi-stage system may be used, where utilisation of the limiting nutrient in the first stage leads to non-repressed conditions in the second stage.

4) Chemostat with biomass feedback is appropriate because the output is increased.

5) Turbidostat mode of operation is appropriate because the fastest growing strain will eventually contribute most to turbidity and thus control the dilution rate of the culture. Slower growing strains will wash out of the vessel because the dilution rate will be above their maximum growth rate on the limiting substrate.

Responses to Chapter 6 SAQs

6.1

1) *Vibrio vulnificus.*

2) *Methanobacterium thermoautotrophicum, Bacillus globisporus.*

3) *Alteromonas haloplanktis.*

4) The doubling times would increase since *Escherichia coli* would no longer be growing at its optimum temperature for growth.

5) False. *Bacillus cereus* has an optimum temperature of 40°C and is therefore a mesophile.

6) *Escherichia coli* and *Neisseria sicca* because their optimum temperature for growth (37°C) is the temperature of the human body.

7) *Alteromonas haloplanktis* (Even if you have not met this species before, its name should have told you it likes salty conditions (halo-) and is planktonic (planktis).

8) *Bacillus globisporus* and *Methanobacterium thermoautotrophicum* are most likely to grow. Although the optimum temperature for growth of *Bacillus cereus* (40°) is closer to 50°C than the optimum temperature for growth of *M. thermoautotrophicum*, the former species is unlikely to grow because 50°C will probably be above its maximum temperature for growth (see Figure 6.2).

9) *Vibrio* is likely to outnumber *Nitrosomonas*. Even though 27°C is closer to the optimum temperature for growth of *Nitrosomonas* than that for *Vibrio*, the data show that *Nitrosomonas* is a much slower growing organism than *Vibrio*.

6.2

1) 105°C.

2) approximately 110°C.

3) approximately 85°C.

4) A temperature range of 85°C to 110°C is most likely since the organism has adapted to grow over this range of temperatures.

5) We can deduce that even though the organism is an extreme thermophile it can survive (but not grow) at low temperature for extended periods of time.

6.3

1) True.

2) False. See Table 6.1.

3) False. Most mutations affecting primary structure of an enzyme decrease its thermal stability.

4) False. These are able to survive but are not able to grow.

5) False. See Figure 6.2.

6) False. Cells may be protected from irreversible damage at freezing temperatures by raising the osmolarity of the cytoplasm.

7) False. Dextran does not penetrate the cell.

6.4 1) True.

2) False. See Table 6.5.

3) True.

4) False. Plasmolysis of cells may occur if the water activity of the cytoplasm is far lower than that of the environment. In this circumstance the cell would lose water by osmosis which could lead to plasmolysis.

5) True.

6) False. See Table 6.5. *Pediococcus halophilus* is a moderate halophile that can grow at 0% NaCl.

7) True.

6.5 Characteristics/processes numbered 2), 3) and 4) would aid an acidophilic bacterium growing at pH 2.0.

Characteristic 1) does not aid the bacterium since cations (protons) would enter the cell and lower the pH of the cytoplasm. It follows that a cell membrane which is relatively impermeable to cations would aid an acidophilic bacterium growing at pH 2.0.

6.6 1) O_2^- - Superoxide free radical (most reactive).

$^1O_2$ Singlet oxygen.

O_2^{2-} - Peroxide.

$^3O_2$ Triplet oxygen (least reactive).

2) Isolate 1 is an *obligate anaerobe*.

Isolate 2 is a *facultative anaerobe*.

Isolate 3 is an *aerotolerant anaerobe* or *obligate anaerobe*.

Isolate 4 is either an *obligate aerobe* or a *microaerophile*.

6.7 1) a) *Moderate barophiles* can be described as *facultative barophiles* because they grow optimally at 400-500 atm but can also grow at 1 atm.

b) *Extreme barophiles* can be described as *obligate barophiles* because they grow at high hydrostatic pressure but cannot grow at atmospheric pressure.

2) Growth rates of barophiles are generally lower than those of non-barophiles because hydrostatic pressure tends to reduce enzyme activity.

3) Moderate barophile. See Figure 6.5. You should note that the organism was able to survive (but not grow) at 1000 atm.

6.8

1) UV

Killing mechanism	Dimer formation in DNA
Wavelength	100-400nm
Protection mechanism	Dark reactivation/photoreactivation

2) Visible light

Killing mechanism	Free radicals/singlet oxygen
Wavelength range	400-700nm
Protection mechanism	Carotenoids

3) X-rays

Killing mechanism	Free radicals
Wavelength range	10^{-4} - 10^{-2} nm
Protection mechanism	Dark reactivation

4) Radiowaves

Killing mechanism	None
Wavelength range	$\sim 10^{10}$ nm
Protection mechanism	None

6.9

Extreme halophile
Xerophile
Facultative psychrophile (psychrotroph)
Osmophile
Acidophile
Microaerophile
Moderate barophile

Responses to Chapter 7 SAQs

7.1

1) Decrease. The high CO_2 levels will tend to produce H_2CO_3 which will dissociate to produce H^+.

2) Increase. Utilisation of amino acids as a carbon source will lead to the production of ammonia. This, in turn, will result in OH^- production from water:

$$NH_3 + H_2O \rightleftarrows NH_4^+ + OH^-$$

3) Probably a decrease. The utilisation of glucose as a carbon source will lead to the production of partially metabolised organic acids. Since the amino acids are only utilised as a nitrogen source, an excess of ammonia will not be generated in this medium.

4) Increase. Due to utilisation (removal) of the acidic carbon source.

5) Increase. Utilisation of nitrate consumes H^+.

7.2   1) a) i) Increased.

ii) Increased.

iii) No change.

b) i) Decreased.

ii) No change.

iii) Increased.

2) a) False. Citric acid is a central metabolite and will therefore be readily utilised by many organisms.

b) False. HPO_4^{2-} will combine with a H^+. It therefore resists a decrease in pH.

c) False. The effective buffering range for phosphates is 6.5 - 8.0.

d) False. In order to minimise the effects on the organism, the buffer should not readily pass through biological membranes.

7.3   1) False. Although partial pressure of oxygen will increase the driving force ($C^* - C_L$) for oxygen transfer, the volumetric transfer coefficient ($k_L a$) will not be altered.

2) True. Small bubbles have a greater interfacial area (a) per unit volume compared to large bubbles. It follows from Equation 7.4 that OTR will therefore increase.

3) True. The solute in culture media generally decreases oxygen concentration relative to that of pure water.

4) False. The concentration of dissolved gas decreases at a constant partial pressure upon the addition of solute. It follows from Henry's law (Equation 7.3) that Henry's constant must increase.

5) False. We can see from Equation 7.2 that OTR is inversely related to Y_o.

6) False. The driving force for oxygen transfer is the difference in oxygen concentration in the two phases.

7.4   1) Sulphite oxidation.

2) Oxygen balance.

3) Dynamic gassing out.

4) Dynamic gassing out.

5) Oxygen balance and sulphite oxidation.

6) Dynamic gassing out.

7.5

1) Stirred tank - flat blade impeller.

2) Loop - propeller type.
Stirred tank - vortex system.
Stirred tank - open inclined blades.

3) All except the stirred tank - vortex type; in which air is dragged into the culture from the air space in the region of the impeller.

4) Stirred tank - open inclined blades.

5) Bubble aeration bioreactor.

6) Bubble aeration bioreactor.
Loop - gas-lift type.
Loop - jet type.

7) Stirred tank - flat blade impeller.

8) Loop - gas-lift type.
Loop - propeller type.
Loop - jet type.

7.6

1) False. A dissolved oxygen sensor is a direct measure of oxygen tension; which, of course, is related to oxygen concentrations according to Henry's law.

2) False. In the galvanic type of dissolved oxygen sensor a small amount of current is drawn from the lead anode to provide a voltage measurement.

3) False. The polarographic type has a platinum cathode and a silver anode.

4) False. The main effect of internal cooling coils is the increase of surface area available for heat transfer.

5) True.

6) True.

7.7

1) Offset is the difference between the new steady-state and the set-point value.

2) Set-point is the desired value for a fermentation parameter.

3) Error signal is related to the difference between the set-point and measured values.

4) Actuator is the final control element.

5) Feed-back control occurs when control action is triggered by an error signal.

6) Direct digital control is where the computer examines an input signal and uses the information to produce an output signal that is sent to the actuator.

7) On-line sensor is a sensor that is part of the fermentation set-up and measures a fermentation parameter without liquid sampling.

Responses to Chapter 8 SAQs

8.1

group a) answers	group b) answers	group c) answers
statements 4	statements 1	statements 2
statements 5	statements 3	statements 7
statements 6	statements 8	

Statements 4 and 5 clearly separate viruses from all living cells. Some structures such as bacterial spores have very limited but definite metabolic activity whereas viruses have absolutely none - thus statement 6 also holds.

There are a very few bacterial species which cannot produce energy or reproduce independently and thus they require host cell participation, for example the mycoplasma group of bacteria. Most cells, as we have indicated, are much larger than 300 nm, a few - again including the mycoplasmas are around 200 nm.

Finally, with reference to statements 2 and 7, all living cells contain nucleic acid and protein.

Viruses are acellular and therefore cannot be considered to be living in the way we define living as applied to cells. If we accept then that viruses are not living entities and have never been living, it is inappropriate to use the term dead - there is an implication of previous life. However viruses are not totally inanimate objects; as indicated earlier, they really are unique!

8.2

1) True. A matter of opinion but probably true in that it was the first way of proving that something other than bacteria could cause disease. Technologically it was certainly the greatest step forward.

2) True. The word filterable was dropped as more became known of viruses. Probably in the early part of the century a lack of knowledge could have allowed the phrase filterable viruses a more wide ranging role than our current concept of viruses.

3) True. Viruses are simple entitles and we think that they arose early in evolutionary terms.

4) False. Bacteriology was still in its infancy and the early work on viruses was carried out on those examples which infect animals and plants.

5) True. Initially some scientists thought viruses consisted of only protein. Quickly however the presence of RNA and its importance in the tobacco mosaic virus were established.

6) False. Viruses are usually very host specific infecting only single species or at most closely related species of cells.

8.3

1) The virus particle which exists outside of the host cell is called a virion.

2) The four major structural groups of viruses are: rod shaped showing helical symmetry; spherical shaped showing icosahedral symmetry; enveloped viruses and complex viruses.

3) The viral nucleic acid is usually surrounded by a protein coat also known as a protein sheath or capsid.

4) The individual protein subunits in the protein coat are called capsomeres.

5) Virus particles which infect bacterial cells are called bacteriophages or phages for short.

6) An icosahedral structure exhibits twenty symmetrical faces.

7) A repeating structure containing two or more different capsomeres is called a morphological unit.

You could have named many possibilities as examples of each of the structural types of viruses; however the ones we mentioned in the text were:

rod shaped helical virus - tobacco mosaic virus;

spherical icosahedral virus - adenovirus;

enveloped virus - herpes virus;

complex virus - T_4 bacteriophage or vaccinia virus (pox virus).

8.4

Recently the International Committee for Taxonomy of Viruses developed a classification system in which viruses were grouped into 50 or so families.

The important properties used to arrive at this separation were:

• nucleic acid types (RNA or DNA);

• nucleic acid strandedness (single or double stranded);

• presence or absence of an envelope.

You would probably have included in your classification scheme (and certainly in an identification scheme):

- capsid symmetry;

- the nature of the host cell - plant, animal or bacterium;

- the type of disease caused.

and possibly:

- the number of morphological units in icosahedral viruses;

- the size of the virion.

8.5   The virion must:

1	firstly locate a suitable host cell	Location
2	attach to or be absorbed onto the host cell	Attachment
3	penetrate the outer layers such that the whole virion (or at least the nucleic acid) is injected into the cytoplasm	Penetration
4	control the metabolic apparatus of the host cell such that new virus material is produced; firstly the viral nucleic acid followed by the capsid and finally, if present other elements such as the envelope	Replication
5	ensure assembly of the new viral components into intact virus particles	Assembly
6	cause release of the new virions into the surroundings, often by lysis of the host cell	Release

8.6

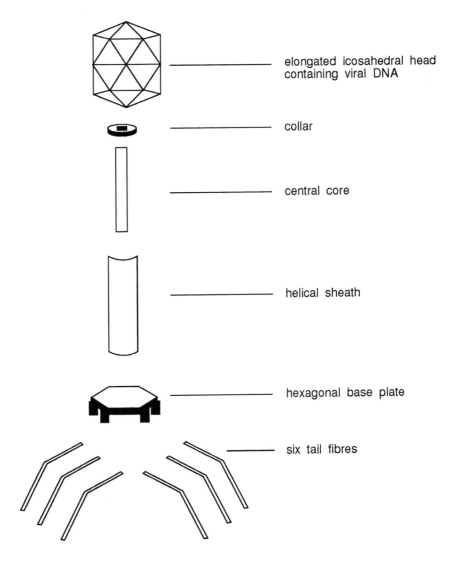

elongated icosahedral head
containing viral DNA

collar

central core

helical sheath

hexagonal base plate

six tail fibres

8.7
1) Receptor sites.

2) Help to penetrate the host cell wall following attachment and to bring about host cell lysis.

3) Releasing viral nucleic acid from the capsid if the entire virion enters the host cell.

4) 20 to 30 minutes (approximately), 300.

5) Virulent.

6) Temperate, prophage.

7) The virus is more likely to survive in times of nutrient shortage for the host cells and will not destroy all the host cells.

8) Induction.

If you managed to get all of these right - well done!

8.8   During the latent period there is no release of infective virions.

In the first part of the latent period - the eclipse phase or period - there are no complete, infective virions within the cells and few if any on the outside. Thus we initially inoculate a culture of bacteria with some virus (the number of PFU's at time 0). As these attach to the hosts' cells and their nucleic acid is injected into the host, then the number of intact viruses declines. Thus after a short period we cannot find any intact virus particles in the culture (the eclipse period).

The rise period is literally that period when the relative plaque forming unit count increases as new virions are released during cell lysis.

The burst size is a measure of the number of virions liberated per infected cell - a number obtained by simple calculation.

8.9   1) The initial number of host cells per ml is:

$71 \times 1/0.1 \times 10^3 = 7.1 \times 10^5$ cells per ml.

10 ml of bacterial suspension were used, thus $7.1 \times 10^5 \times 10$ cells = 7.1×10^6 cells.

2) There were 10^7 beads per ml to start with.

The mixture of beads : virions was equal parts.

The ratio of beads : virions was 5 to 19.

Therefore there were $19/5 \times 10^7$ virions = 3.8×10^7 virions per ml.

100 ml were used thus the total number of virions was 3.8×10^9.

3) From the plaque assay data:

21 plaques from 0.1 ml of a 10^{-5} dilution.

Therefore there were $21 \times 10 \times 10^5$ plaques = 2.1×10^7 plaques per ml.

100 ml were used thus the total number of plaques was 2.1×10^9.

4) The efficiency of plating relates the virion number determined by electron microscopy to the number of plaques formed. Thus:

the percentage efficiency is $\dfrac{2.1 \times 10^9}{3.8 \times 10^9} \times 100 = 55.2\%$.

5) Using the plaque method there were 2.1×10^9 virions released from 7.1×10^6 cells, therefore $2100/7.1 = 296$ virions per host cell.

   Using the electron microscopy method, there were $3800/7.1 = 535$ virions per host cell.

In practice it is impossible to arrange it such that all host cells are infected by a single virion. If we wish to use the plaque count method to determine the number of viable phage particles we usually conduct the experiment with a vast excess of host cells. If, on the other hand, we wish to determine 'burst size' (ie the number of virions released per infected host) we usually infect a host with a known number of PFU's and then measure the number of PFU's after lyses has occurred. The burst size will then be given by

$$\frac{\text{Final PFU count} - \text{Initial PFU count}}{\text{Initial PFU count}}$$

Responses to Chapter 9 SAQs

9.1

1) The D_{10}-value can be determined from the slope of line B.

   eg: At $\log_{10}$ number of survivors = 2, Time = 1.8 h

   At $\log_{10}$ number of survivors = 1, Time = 4.0 h

   $\log_{10}2 - \log_{10}1 = 10$-fold decrease.

   Therefore D_{10}-value = $4.0 - 1.8 = 2.2$ h.

2) Factor b) increases the rate of killing by a disinfectant.

   You might have also included c - but think carefully about this. Although increasing time would increase the number of cells killed, it would not increase the rate at which the cells were killed.

9.2

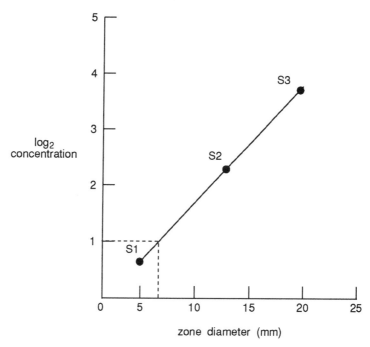

Zone diameter solution Y = 6.8 mm

From the graph, 6.8 mm = $\log_2$ concentration Y = 1

Antilog $\log_2$ 1 = 2

Antimicrobial concentration of solution Y = $2\,\mu g\ ml^{-1}$.

You should note that log concentration is plotted against zone diameter to obtain a linear relationship; this reduces possible errors associated with determination of concentration for an 'unknown' solution.

9.3   1) Phenol has characteristic c) and g).

2) Chlorine has characteristics a), d) and h).

3) Ethanol has characteristics b), c) and g).

4) Cationic detergents have characteristics a), b) and e).

5) Ethylene oxide has characteristics f) and h).

9.4   1) Tetracycline, ampicillin.

2) All antibiotics listed have a degree of selectivity.

3) Pencillin V, oxacillin, ampicillin.

4) Penicillin V, oxacillin, ampiciliin.

5) Polymixin.

6) Penicillin V.

7) Cycloheximide.

8) Streptomycin.

9) Streptomycin.

10) Chloramphenicol.

9.5

1) False. Suphanilamide is an analogue of p-amino benzoic acid which is an intermediate in the synthesis of folic acid.

2) False. 5-Bromouracil is an analogue of the DNA base thymine. 5-Fluorouracil is an analogue of uracil.

3) True. Phenylalanine will complete with its analogue p-fluorophenylalanine in biosynthetic pathways.

4) True.

5) True. By reducing toxicity.

6) False. Penicillin G is inactivated by stomach acids.

9.6

1) True.

2) False. Methicillin is a penicillin derivative that is resistant to penicillinase (β-lactamase).

3) False. R factors transfer antibiotic resistance only between closely related bacteria.

4) False. Bacterial endospores are naturally resistant to certain disinfectants.

Appendix 1 - Categorisation of micro-organisms according to hazard

Within the main body of the text, we have referred to the fact that micro-organisms may be categorised according to the extent that they represent a hazard to health. Four groups are identified - hazard group 1 being of lowest risk, hazard group 4 presenting the greatest hazard.

The extent and nature of the containment that needs to be implemented when using micro-organisms is governed to a large extent by the hazard group to which the micro-organism belongs.

Here we provide a list of organisms according to their hazard group. The list includes only organisms in hazard groups 2, 3 and 4. The names that have been used are those in common (but not universal) usage. Many have one or more synonyms.

The list provided here is based on the categorisation of pathogens by the UK, Advisory Committee on Pathogens ('Categorisation of Pathogens according to hazard and categories of containment' HMSO, 1990).

It countries other than the UK, slightly different nomenclature may be used although the same (or very similar) criteria for assigning micro-organisms to 1 of 4 categories is often employed.

The list provided here should not be regarded as universally applicable for all time and for all places. As new knowledge emerges or as new control measures are developed, it becomes appropriate to re-categorise some organism. Furthermore, it is possible to develop attenuated or genetically modify strains of pathogens organism such that they may be re-categorised.

The reader should also be aware that specific conditions may apply to particular pathogens. For example, deliberate cultivation of the Variola (smallpox) virus is banned totally in some countries. A key aspect of working with micro-organisms is therefore, not only to work safely, but also be aware of the regulatory obligations that must be fulfilled.

BACTERIA, CHLAMYDIAS, RICKETTSIAS AND MYCOPLASMAS

Hazard Group 3

Bacillus anthracis

Brucella spp.

Chlamydia psittaci (avian strains only)

Coxiella burnetti

Francisella tularensis (Type A)

Mycobacterium africanum

Mycobacterium avium

Mycobacterium bovis (excl BCG strain)

Mycobacterium intracellulare

Mycobacterium kansasii

Mycobacterium leprae

Mycobacterium malmoense

Mycobacterium paratuberculosis

Mycobacterium scrofulaceum

Mycobacterium simiae

Mycobacterium szulgai

Mycobacterium tuberculosis

Mycobacterium xenopi

Pseudomonas mallei

Pseudomonas pseudomallei

Rickettsia-like organisms

Rickettsia spp.

Salmonella paratyphi A, B, C

Salmonella typhi

Shigella dysenteriae (Type 1)

Yersinia pseudotuberculosis subsp pestis (*Y pestis*)

Hazard Group 2

Acinetobacter calcoaceticus

Acintoebacter lwoffi

Actinobacillus spp.

Actinomadura spp.

Actinomyces bovis

Actinomyces israelii

Aeromonas hydrophila

Alcaligenes spp.

Arizona spp.

Bacillus cereus

Bacteroides spp.

Bacterionemia matruchottii

Bartonella bacilliformis

Bordetella parapertussis

Bordetella pertusis

Borrelia spp.

Campylobacter spp.

Cardiobacterium hominis

Chlamydia spp. (other than aivan strains)

Clostridium botulinum

Legionella spp.

Leptospira spp.

Listeria monocytogenes

Moraxella spp.

Morganella morganii

Mycobacterium bovis (BCG strain)

Mycobacterium chelonei

Mycobacterium fortuitum

Mycobacterium marinum

Mycobacterium microti

Mycobacterium ulcerans

Mycoplasma pneumoniae

Neisseria spp. (spp .known to be pathogenic for man)

Nocardia asteroides

Nocardia brasiliensis

Pasteurella spp.

Peptostreptococcus spp.

Plesiomonas shigelloides

Proteus spp.

Providencia spp.

Hazard Group 2 (Cont)

Clostridium tetani

Clostridium spp. (other spp. known to be pathogenic for man)

Corynebacterium diphtheriae

Corynebacterium spp. (other spp. known to be pathogenic for man)

Edwardsiella tarda

Eikenella corrodens

Enterobacter spp.

Erysipelothrix rhusiopathiae

Escherichia coli (except those known to be non-pathogenic)

Flavobacterium meningosepticum

Francisella tularensis (Type B)

Fusobacterium spp.

Gardnerella vaginalis

Haemophilus spp.

Hafnia alvei

Kingella kingae

Klebsiella spp.

Pseudomonas spp. (other spp. known to be pathogenic for man)

Salmonella spp. (other than those in Hazard Group 3)

Serratia liquefaciens

Serratia marcescens

Shigella spp. (other than that in Hazard Group 3)

Staphylococcus aureus

Streptobacillus moniliformis

Streptococcus spp. (except those known to be non-pathogenic for man)

Treponema pallidum

Treponema pertenue

Veillonella spp.

Vibrio cholerae (incl El Tor)

Vibrio parahaemolyticus

Vibrio spp. (other species known to be pathogenic for man)

Yersinia enterocolitica

Yersinia pseudotuberculosis subsp pseudotuberculosis

FUNGI

Hazard Group 3

Blastomyces dermatitidis

(Ajellomyces dermatitidis)

Coccidioides immitis

Histoplasma capsulatum var. capsulatum

(Ajellomyces capsulata)

Histoplasma capsulatum var duboisii

Histoplasma capsulatum var farciminosum

Paracoccidioides brasiliensis

Penicillium marneffei

Hazard Group 2

Absidia corymbifera

Acremonium falciforme

Acremonium kiliense

Acremonium recifei

Aspergillus flavus

Aspergillus fumigatus

Aspergillus nidulans

Exophialia jeanselmei

Exophialia spinifera

Exophialia richardsiae

Fonsecaea compacta

Fonsecaea pedrosoi

Fusarium solani

Fusarium oxysporum

Hazard Group 2 (Cont)

Aspergillus niger

Aspergillus terreus

Basidiobolus haptosporus

Candida albicans

Candida glabrata

Candida guilliermondii

Candida krusei

Candida parapsilosis

Candida kefyr

Candida tropicalis

Cladosporium carrionii

Conidiobolus coronatus

Cryptococcus neoformans

(Filobasidiella neoformans)

Cunninghamella elegans

Curvularia lunata

Emmonsia parva

Emmonsia parva var. crescens

Epidermophyton floccosum

Exophialia dermitidis

Exophialia werneckii

Geotrichum candidum

Hendersonula toruloidea

Leptosphaeria senegalensis

Madurella mycetomatis

Madurella grisea

Malassezia furfur

Microsporum spp.

Neotestudina rosatii

Phialophora verrucosa

Piedraia hortae

Pneumocytis carinii

Pseudallescheria boydii

Pyrenochaeta romeroi

Rhizomucor pusillus

Rhizopus microsporus

Rhizopus oryzae

Sporothrix schenckii

Trichophyton spp.

Trichosporon beigelii

Xylohypha bantiana

PARASITES

Hazard Group 3

Echinococcus spp.

Leishmania spp. (mammalian)

Naegleria spp.

Toxoplasma gondii

Trypanosoma cruzi

Hazard Group 2

Acanthamoeba spp.

Ancylostoma duodenale

Angiostrongylus spp.

Ascaris lumbricoides

Babesia microti

Babesia divergens

Balantidium coli

Brugia spp.

Capillaria spp.

Loa loa

Mansonella ozzardi

Necator americanus

Onchocerca volvulus

Opisthorchis spp.

Paragonimus westermanni

Plasmodium spp. (human & simian)

Pneumocystis carinii

Schistosoma haematobium

Hazard Group 2 (Cont)

Clonorchis sinensis	*Schistosoma intercalatum*
Cryptosporidium spp.	*Schistosoma japonicum*
Dipetalonema streptocerca	*Schistosoma mansoni*
Dipetalonema perstans	*Stronglyloides spp.*
Diphyllobothrium latum	*Taenia saginata*
Drancunculus medinensis	*Taenia solium*
Entamoeba histolytica	*Toxocara canis*
Fasciola hepatica	*Trichinella spp.*
Fasciola gigantea	*Trichomonas vaginalis*
Fasciolopsis buski	*Trichostrongylus spp.*
Giardia lamblia	*Trichuris trichiura*
Hymenolepis nana (human origin)	*Trypanosoma* brucei subsp
Hymenolepis diminuta	*Wuchereria bancrofttii*

VIRUSES

Hazard Group 4

Arenaviridae

Junin virus	Machupo virus
Lassa fever virus	Mopeia virus

Bunyaviridae **Togaviridae**

Nairoviruses	Flaviviruses
Congo/Crimean haemorrhagic fever	Tick-borne viruses

Filoviridae

Ebola virus	Absettarov
Marburg virus	Hanzalova
	Hypr

Poxviridae

Variola (major & minor) virus	Kyasanur Forest
('whitepox' virus)	Omsk
	Russian spring-summer encephalitis

Hazard Group 3

Arenaviridae **Rhabdoviridae**

Lymphocytic choriomeningitis virus (LCM)	Rabies virus

Hazard Group 3 (Cont)

Bunyaviridae

Bunyamwera supergroup

Oropouche virus

Phleboviruses

Rift Valley fever

Hantaviruses

Hantaan (Korean haemorrhagic fever)

Other hantaviruses

Hepadnaviridae

Hepatitis B virus

Hepatitis B virus + Delta

Herpesviridae

Herpesvirus simiae (B virus)

Poxviridae

Monkeypox virus

Retroviridae

Human immunodeficiency viruses (HIV)

Human T-cell lymphotropic viruses (HTLV) types 1 and 2

Hazard Group 2

Adenoviridae

Arenaviridae

other arenaviruses

Astroviridae

Bunyaviridae

Hazara virus

other bunyaviruses

Caliciviridae

Coronaviridae

Herpesviridae

Cytomegalovirus

Epstein-Barr virus

Herpes simplex viruses types 1 and 2

Herpesvirus varicella-zoster

Human B-lymphotropic virus (HBLV - human herpesvirus type 6)

Togaviridae

Alphaviruses

Eastern equine encephalomyelitis

Venezuelan equine encephalomyelitis

Western equine encephalomyelitis

Flaviviruses

Japanese B encephalitis

Kumlinge

Louping ill

Murray Valley encephalitis (Australia encephalitis)

Powassan

Rocio

St Louis encephalitis

Tick-borne encephalitis

Yellow fever

Picornaviridae

Acute haemorrhagic conjunctivitis virus (AHC)

Coxsackieviruses

Echoviruses

Hepatitis A virus (human enterovirus type 72)

Polioviruses

Rhinoviruses

Poxviridae

Cowpox virus

Molluscum contagiosum virus

Orf virus

Vaccinia virus

Reoviridae

Human rotaviruses

Orbiviruses

Reoviruses

Orthomyxoviridae

Influenza viruses types A, B & C

Influenza virus type A-recent isolates

Paramyxoviridae

Measles virus

Mumps virus

Newcastle disease virus

Parainfluenza viruses types 1 to 4

Respiratory syncytial virus

Papovaviridae

BK and JC viruses

Human papillomaviruses

Parvoviridae

Human parvovirus (B19)

Rhabdoviridae

Vesicular stomatitis virus

Togaviridae

Other alphaviruses

Other flaviviruses

Rubivirus (rubella)

Unclassified viruses

Hepatitis non-A non-B viruses

Norwalk-like group of small round structured viruses

Small round viruses (SRV - associated with gastroenteritis)

Unconventional agents associated with:

Creutzfeldt-Jakob disease

Gertsmann-Sträussler-Schienker syndrome

Kuru

Index

A

acceleration phase, 86
accuracy, 54
ACDP, 13
acetate, 75 , 131
acetic acid, 162
acidophiles, 161
    facultative, 162
    obligate, 162
acquired resistance, 274
active transport, 153
acute toxicity, 273
adenosine triphosphate (ATP), 69
adenovirus, 221
aeration rates, 193
Aerobacter aerogenes, 95
aerobes, 163
aerobic conditions, 42
aerobic respiration, 37 , 163
aerosols, 158
aerotolerant anaerobes, 163
agar, 10 , 33
agar diffusion method, 250
agar overlay, 238
agglutination, 239
air bubbles, 200
airborne infections, 165
alcohols as antiseptics, 255
algae, 4 , 42 , 58 , 87 , 148 , 245
algistatic, 245
alkalophiles, 161
allergic reaction, 274
Alteromonas haloplanktis, 159
amino penicillanic acid, 260
aminobenzoic acid, 270
aminoglycoside antibiotics, 267
aminoglycosides, 275
ammonia, 162 , 182
ammonia consumption, 73
ammonia toxicity, 182
ammonium, 42
ammonium sulphate precipitation, 240
Amphotericin B, 261
ampicillin, 264 , 265
anabolism, 27
anaerobes
    aerotolerant, 163
anaerobic conditions, 42
anaerobic respiration, 37
animal models, 274
animal viruses, 221 , 238 , 239

Animalia, 4
antibacterial agents, 236
antibiotic resistance, 274
antibiotic toxicity, 273
antibiotics, 244 , 247 , 259 , 260
antibiotics in agriculture, 276
antifungal agents, 236
antimicrobial activity
    measurement of, 247
antimicrobial agents, 244
    effects of, 245
    selectivity of, 259
antiseptics, 245 , 253
APA, 260
Aquaspirillum serpens, 159
Arrhenius equation, 33
aseptic conditions, 141
aseptic technique, 40
Aspergillus niger, 120 , 186
asynchronous growth, 81
ATP, 93
attachment of viruses, 226
attenuation, 10
autoclave, 49
autolysins, 262
automatic control, 183
autotrophs, 25 , 42
axenic culture, 30
axial flow, 195
azidothymidine (AZT), 271
Azotobacter, 10 , 43 , 167
aztreonam, 276

B

Bacillus, 44 , 49
Bacillus anthracis, 8
Bacillus polymyxa, 261 , 266
Bacillus subtilis, 261
Bacitracin, 261
bacteria, 4 , 35 , 42 , 49 , 148 , 188 , 245
    effect of pH, 35
bacterial cell cycle, 96
bacterial cells, 255
bacterial meningitis, 267
bacteriocidal, 246
bacteriolytic, 246
bacteriophage, 218 , 222 , 227 , 236 , 238 , 241
bacteriostatic, 245 , 246
bacteriostatic antibiotic, 267
baffles, 194
baker's yeast, 128
balanced growth, 27 , 81
barophiles, 39 , 168